The Smart Grid:
Enabling Energy Efficiency and Demand Response

The Smart Grid:
Enabling Energy Efficiency and Demand Response

Clark W. Gellings, P.E.

THE FAIRMONT PRESS, INC.

CRC Press
Taylor & Francis Group

Library of Congress Cataloging-in-Publication Data

Gellings, Clark W.
   The smart grid : enabling energy efficiency and demand response / Clark W.
Gellings.
      p. cm.
   Includes bibliographical references and index.
   ISBN-10: 0-88173-623-6 (alk. paper)
   ISBN-10: 0-88173-624-4 (electronic)
   ISBN-13: 978-1-4398-1574-8 (Taylor & Francis distribution : alk. paper)
     1. Electric power distribution--Energy conservation. 2. Electric power--
Conservation. 3. Electric utilities--Energy conservation. I. Title.

   TK3091.G448 2009
   621.319--dc22

                                                2009013089

Published by The Fairmont Press, Inc.
700 Indian Trail
Lilburn, GA 30047
tel: 770-925-9388; fax: 770-381-9865
http://www.fairmontpress.com

Distributed by Taylor & Francis Ltd.
6000 Broken Sound Parkway NW, Suite 300
Boca Raton, FL 33487, USA
E-mail: orders@crcpress.com

Distributed by Taylor & Francis Ltd.
23-25 Blades Court
Deodar Road
London SW15 2NU, UK
E-mail: uk.tandf@thomsonpublishingservices.co.uk

Printed in the United States of America
10 9 8 7 6 5 4 3 2 1

10: 0-88173-623-6 (The Fairmont Press, Inc.)
13: 978-1-4398-1574-8 (Taylor & Francis Ltd.)

While every effort is made to provide dependable information, the publisher,
authors, and editors cannot be held responsible for any errors or omissions.

Contents

Definitions

ADA	Advanced Distribution Automation
ANSI	American National Standards Institute
ASDs	Adjustable Speed Drives
CCS	Carbon Capture and Storage
CEIDS	The Consortium for Electric Infrastructure to Support a Digital Society
CFL	Compact Fluorescent Lamps
CH_4	Methane
CHP	Combined Heat and Power
CO_2	Carbon Dioxide
CPP	Critical Peak Period
CVR	Conservation Voltage Reduction
DA	Distribution Automation
DC	Direct Current
DER	Distributed Energy Resources
DG	Distributed Generation
DOE	U.S. Department of Energy
DR	Demand Response
DSE	Distribution System Efficiency
DSM	Demand-side Management
E2I	Electricity Innovation Institute
EDF	Electricite de France
EMCS	Energy Management Control System
EMS	Energy Management System
EPRI	Electric Power Research Institute
FAC	TAC Transmission Systems
FD	Forced Draft
FERC	Federal Energy Regulatory Commission
FSM	Fast Simulation and Modeling
GE	General Electric
GT	Gas Turbine
GTO-PWM	Gate Turn-Off Thyrister Pulse-width Modulated
HLA	High-Level Architecture
HP	Horsepower
HPS	High-Pressure Sodium

HPWHs	Heat Pump Water Heaters
HRSG	Heat Recovery Steam Generator
HVAC	Heating, Ventilation and Air Conditioning
ID	Induced Draft
IEA	International Energy Agency
IECSA	Integrated Energy and Communications Architecture
IEEE	Institute of Electrical and Electronics Engineers
INAs	Intelligent Network Agents
IP	Internet Protocol
ISOs	Independent System Operators
IT	Information Technology
IUT	Intelligent Universal Transformer
KHz	Kilohertz
kWh	Kilowatt hour
LCI	Load Commutated Inverter
LED	Light-Emitting Diode
LEDSAL	Light-Emitting Diode Street and Area Lighting
MGS	Modern Grid Strategy
MH	Metal Halide
MHz	Megahertz
MW	Megawatt
MWh	Megawatt Hour
NAAQS	National Ambient Air Quality Standards
NEMA	National Electrical Manufacturers Association
NERC	North American Electric Reliability Council
NETL	National Energy Technology Laboratory
NSR	New Source Review
PDAs	Personal Digital Assistants
PHEVs	Plug-In Hybrid Electric Vehicles
PMUs	Phaser Measurement Units
PV	Photovoltaics
RF	Radio Frequency
ROI	Return on Investment
RTOs	Regional Transmission Organizations
SHG	Self-Healing Grid
SMPS	Switched Mode Power Supply
SQRA	Security Quality Reliability and Availability
TES	Thermal Energy Storage
TCP/IP	Transmission Control Protocol/Internet Protocol

TOU	Time of Use
TWh	Terawatt Hour
UML	Unified Modeling Language
UPS	Uninterruptible Power Supply
V	Voltage
VFD	Variable Frequency Drive
BRF	Variable Refrigerant Flow
WAMs	Wide Area Monitoring System
WEO	World Energy Outlook

Chapter 1

What is the Smart Grid?

WHAT IS A SMART GRID?

The electric power system delivery has often been cited as the greatest and most complex machine ever built. It consists of wires, cables, towers, transformers and circuit breakers—all bolted together in some fashion. Sometime in the 1960s, the industry initiated the use of computers to monitor and offer some control of the power system. This, coupled with a modest use of sensors, has increased over time. It still remains less than ideal—*for example*, power system area operators can, at best, see the condition of the power system with a 20-second delay. Industry suppliers refer to this as "real time." However, 20 seconds is still not real time when one considers that the electromagnetic pulse moves at nearly the speed of light.

Actually, the electric power delivery system is almost entirely a mechanical system, with only modest use of sensors, minimal electronic communication and almost no electronic control. In the last 25 years, almost all other industries in the western world have modernized themselves with the use of sensors, communications and computational ability. For these other industries, there has been enormous improvements in productivity, efficiency, quality of products and services, and environmental performance.

In brief, a smart grid is the use of sensors, communications, computational ability and control in some form to enhance the overall functionality of the electric power delivery system. A dumb system becomes smart by sensing, communicating, applying intelligence, exercising control and through feedback, continually adjusting. For a power system, this permits several functions which allow optimization—in combination—of the use of bulk generation and storage, transmission, distribution, distributed resources and consumer end uses toward goals which ensure reliability and optimize or minimize the use of energy, mitigate environmental impact, manage assets, and contain cost.

THE SMART GRID ENABLES THE ELECTRINET[SM]

The ElectriNet[SM] is the guiding concept for marrying the smart grid with low-carbon central generation, local energy networks and electric transportation (see Figure 1). The ElectriNet[SM] recognizes the evolution of the power system into a highly interconnected, complex, and interactive network of power systems, telecommunications, the Internet, and electronic commerce applications. At the same time, the move towards more competitive electricity markets requires a much more sophisticated infrastructure for supporting myriad informational, financial, and physical transactions between the several members of the electricity value chain that supplement or replaces the vertically integrated utility. This next-generation electrical infrastructure, the ElectriNet[SM], will provide for seamless integration/interoperability of the many disparate systems and components, as well as enable the ability to manage competitive transactions resulting from competitive service offerings that emerge in the restructured utility environment. Examples of competitive transactions include settlements for demand response participation, information reporting and notification, energy trading, and bidding capacity into ancillary service markets.

Realizing the ElectriNet[SM] depends on developing the IntelliGrid[SM] (see Chapter 6) communications architecture to enable connectivity between each element of the ElectriNet[SM] with requirements for developing agent-based software systems, which can facilitate the informational, financial, and physical transactions necessary to assure

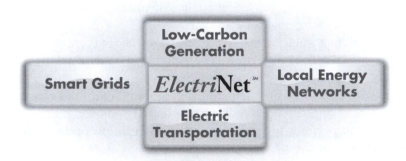

Creating the Electricity Network of the Future

Figure 1-1. Action Framework—Four Evolving Infrastructures

adequate reliability, efficiency, security, and stability of power systems operating in competitive electricity markets. In addition, the architectural requirements will be designed to support multiple operational criteria, including analysis and response to electrical grid contingencies, pricing, and other market/system conditions. The goals of the architecture are to allow for interoperability and flexibility to facilitate and enable competitive transactions to occur. Interoperability can be enabled by the use of open communication protocols. Flexibility can be provided by the specification of user-defined business rules which capture the unique needs of various service offerings.

The ElectriNet^{SM} provides a new perspective on how to manage transactions given the nature of the existing and emerging distributed, heterogeneous communications and control network. This perspective is based on combining distributed computing technologies such as web services, the semantic web, and intelligent agents. Developing an architecture allows future developers to access this framework as a resource or design pattern for developing distributed software applications, taking into account the core concepts of interoperability and support for multiple operational criteria (business rules). The purpose of this architecture is to provide a resource that can serve as a road map to understanding, applying and building next-generation agent-based software systems applied to electricity value chain transactional systems.

A new energy value chain is emerging as a result of new technologies, new players, and new regulatory environments that encourage competitive markets. In the case of electricity, the value chain proceeds as follows: It starts at the fuel/energy source; proceeds to the power generator; continues when the energy is delivered through the high-voltage transmission networks; continues when the electricity is stepped down to a lower voltage onto medium-voltage distribution networks; and finally is delivered to end-use customers for consumption. There are a large number of operational services along this value chain for delivering electricity to customers. Much of the existing focus has been on the supply side to enable competitive wholesale transactions resulting in trading floors for energy and capacity sales, as well as promoting open and non-discriminatory access to the transmission grid. In this new energy value chain, the consumption or demand side of electricity deserves special attention.

Changes in technology and the resulting economics have now disrupted that traditional value chain and stimulated the adoption of distributed energy resources (DER). These distribution resources can take many forms, but some key examples are distributed generation and storage and plug-in hybrid electric vehicles (PHEVs). In addition, because of competition and deregulation, a whole new area of energy services and transactions has sprung up around the demand side of the value chain. One of these new energy services is demand response (DR) which enables load and other DER resources to provide capacity into the bulk power system in response to grid contingencies and market pricing signals. DR is an example of an energy service which requires the interaction and integration of multiple-party business systems and physical assets resulting in both physical and financial transactions. A whole class of service offerings which have similar requirements include those related to customer billing, management of customer equipment, energy information, and a range of value-added services are emerging (e.g., on-line meter reading, bill management, energy audits, energy information, real-time pricing, procurement, etc). Many of these service offerings share similar requirements for integrating disparate systems, automating business processes, and enabling physical and financial transactions. Delivering these services will require a communications architecture that is open, highly scalable, and sufficiently flexible and adaptable to meet the changing business needs of suppliers and customers.

The proposed architecture is designed to enable communication and decision making between distributed system nodes and parties. The architecture is intended to be used to develop software that can supplement the existing power distribution/market network communication infrastructure. It could consist of a collection of reusable software agents and associated hardware specifications that will interoperate within the many interfaces and devices on the power system/market infrastructure. The use of software agents allows the ability for communication and cooperation among system nodes, while taking into account the specific business and technical requirements of many industry players such as energy users, distribution companies, transmission companies, energy service companies, and energy market operators.

LOCAL ENERGY NETWORKS

The local energy network facilitates the functionality of the ElectriNet[SM]. Overall, the combination allows for the operation of a power system that is self-sensing, secure, self-correcting and self-healing and is able to sustain failure of individual components without interrupting service. As such, it is able to meet consumer needs at a reasonable cost with minimal resource utilization and minimal environmental impact and, therefore, enhance the quality of life and improve economic productivity

Local energy networks increase the independence, flexibility and intelligence for optimization of energy use and energy management at the local level; and then integrate local energy networks to the smart grid. Local energy networks, energy sources and a power distribution infrastructure are integrated at the local level. This could be an industrial facility, a commercial building, a campus of buildings, or a residential neighborhood. Local area networks are interconnected with different localized systems to take advantage of power generation and storage through the smart grid enabling complete integration of the power system across wide areas. Localized energy networks can accommodate increasing consumer demands for independence, convenience, appearance, environmentally friendly service and cost control.

ELECTRIC TRANSPORTATION

The next building block of the ElectriNet[SM] is electric transportation—particularly electric vehicles and, in the near-term, plug-in hybrid electric vehicles (PHEVs). As PHEVs begin to proliferate, the availability of both a controllable load and controllable on-site electrical storage can have a profound impact on the electrical systems.

Plug-in hybrid electric vehicles represent the most promising approach to introducing the significant use of electricity as transportation fuel.

PHEV development can build on more than a decade of experience with conventional hybrids such as the Toyota Prius and Ford Escape, which use a battery and electric motor to augment the power of an internal combustion engine. To this blend of technologies, PHEVs add the ability to charge the battery using low-cost, off-peak electricity from the grid—allowing a vehicle to run on the equivalent of 75¢ per gallon or better at today's electricity price. And PHEVs draw only about 1.4-2 kW of power while charging—about what a dishwasher draws.

The primary challenges to widespread use of PHEVs, challenges that will require direct utility involvement to overcome, include specification of the local energy network and development of a mass market to lower battery costs.

LOW-CARBON CENTRAL GENERATION

An essential element of the ElectriNet[SM] is low-carbon central generation. The complete integration of the power system across wide areas must include the availability of central generation and large-scale central storage. The ElectriNet[SM] facilitates the inclusion of multiple centralized generation sources linked through high-voltage networks. The design implies full flexibility to transport power over long distances to optimize generation resources and the ability to deliver the power to load centers in the most efficient manner possible coupled with the strong backbone.

Successful implementation of the ElectriNet[SM] assumes successful achievement of performance and deployment targets associated with several advanced technologies as a basis for estimating CO_2 emissions reduction potential. Those related to central generation must include an expanded use of renewable energy, particularly wind, solar-thermal, solar-photovoltaics (PV) and biomass, continued use of the existing nuclear fleet through life extension, as well as deployment of advanced light water nuclear reactors, advanced coal power plants operating at substantially higher temperatures and pressures, and wide-scale use of CO_2 capture and storage after 2020.

WHAT SHOULD BE THE ATTRIBUTES OF THE SMART GRID?

In order for the smart grid to enable the ElectriNet[SM], the following attributes would need to be addressed:

- Absolute reliability of supply.

- Optimal use of bulk power generation and storage in combination with distributed resources and controllable/dispatchable consumer loads to assure lowest cost.

- Minimal environmental impact of electricity production and delivery.

- Reduction in electricity used in the generation of electricity and an increase in the efficiency of the power delivery system and in the efficiency and effectiveness of end uses.

- Resiliency of supply and delivery from physical and cyber attacks and major natural phenomena (e.g., hurricanes, earthquakes, tsunamis, etc.).

- Assuring optimal power quality for all consumers who require it.

- Monitoring of all critical components of the power system to enable automated maintenance and outage prevention.

WHY DO WE NEED A SMART GRID?

The nation's power delivery system is being stressed in new ways for which it was not designed. For example, while there may have been specific operational, maintenance and performance issues that contributed to the August 14, 2003 outage, a number of improvements to the system could minimize the potential threat and severity of any future outages.

The original design of the power delivery system renders some areas of the United States particularly vulnerable. For example, the North American power delivery system was laid out in cohesive local electrical zones. Power plants were located so as to serve the utility's local residential, commercial, and industrial consumers. Under deregulation of wholesale power transactions, electricity generators, both traditional utilities and independent power producers, were encouraged to transfer electricity outside of the original service areas to respond to market needs and opportunities. This can stress the transmission system far beyond the limits for which it was designed and built. These constraints can be resolved but they require investment and innovation in the use of power delivery technologies.

The U.S. delivery system (transmission and distribution) is largely based upon technology developed in the 1940s and 1950s and installed over the next 30 to 50 years. In the period from 1988 to 1998, electricity demand in the U.S. grew by 30%, yet only 15% of new transmission capacity was added. According to a North American Electric Reliability Council (NERC) reliability assessment, demand is expected to grow

20% during the 10 years from 2002 to 2011 while less than 5% in new transmission capacity is planned. Meanwhile, the number of wholesale transactions each day has grown by roughly 400% since 1998. This has resulted in significantly increased transmission congestion—effectively, bottlenecks in the flow of wholesale power—which increases the level of stress on the system.

A lot has been done to mitigate the potential for blackouts—particularly in the effort to provide new technologies that can help make electricity more reliable, in order to sustain an increasingly high-tech economy which is based, in part, on the use of power-sensitive equipment. Many of these technologies are ready for wide deployment now, while others are only now entering demonstrations.

In addition, the nation is increasingly embracing the use of renewable power generation. However, large wind and solar resources are located far from population centers. Substantial new transmission must be built to add these resources to the nation's generation portfolio.

Expanding transmission and applying new technologies will require a great deal of cooperation between government and industry. Specifically, six steps should be taken to accelerate the formation of a smart grid and to enable the expansion of renewable generation and to reduce the risk of having regional blackouts—which will surely come if these steps aren't taken.

1. Build new generation and transmission facilities in coordination with each other, on a regional basis. Without such coordination, generators will be tempted to build new plants where local prices are high—and then oppose construction of new power lines that could bring in cheaper power. Regional transmission organizations (RTOs) and independent system operators (ISOs) should be given the responsibility and necessary authority to carry out a program of coordinated expansion planning.

2. Implement technologies necessary for wide-area grid operations. For an RTO or ISO to operate a large regional power system, key element of the grid must be "observable"—either through direct monitoring or computerized estimation. Wide-area monitoring systems (WAMS) employing phaser measurement units (PMUs) are now being applied to provide direct measurements. In addition, the use of advanced software for state estimation (modeling

the probable status of grid elements that aren't monitored directly) should become mandatory. So should the expanded use of security assessment software, which can help operators mitigate problems when they arise.

3. The common practice of operating power systems under conditions (so-called "N-1 contingencies") that security assessment software indicates might lead to blackouts should be reconsidered, as should the type of operations planning that sets up such conditions. Some areas have already employed more constrained operation, such as New York State.

4. Grid operations must be coordinated more closely with power market operations, in order not only to minimize the risk of more blackouts but also prevent the type of price spikes experienced already in areas like California. Specifically, prices must be determined according to market rules established to ensure that power flows are handled more cost-efficiently and transmission congestion is avoided. Consumers also need to be provided with ways to curb demand automatically, as needed, in return for price breaks.

5. Improve emergency operations. Clear lines of authority are needed to handle emergencies effectively. System operators also need to be trained more thoroughly in grid restoration and "black starts" (bootstrapping recovery after a blackout). The whole question of how to set protective relays in order to prevent the "cascade" of an outage across a whole region also needs to be reexamined.

6. Information systems and procedures need to be updated. Complex data communications underlies power system operations, especially during an emergency. Many of these systems need upgrading, using advanced technologies, and the procedures for their use should also be fundamentally revised.

The electric power industry has long presented to the world a gold standard of reliability in power system operations. Problems over time provide a warning that this standard will be tarnished unless steps are taken to ensure even higher levels of reliability for the future. Otherwise, the costs associated with poor system reliability could significantly damage the world economy as a whole.

In a digital age when consumers demand higher quality, more reliable power and long-distance power trades place unprecedented demands on the system, adequate investment in the nation's electric infrastructure is critical. The development and deployment of a more robust, functional and resilient power delivery system is needed. The overall system is being called a smart grid. Under this definition, the smart grid is an advanced system that will increase the productivity resulting from the use of electricity, and at the same time, create the backbone application of new technologies far into the future.

This conceptual design of the smart grid addresses five functionalities which should be part of the power system of tomorrow. These functionalities are as follows:

Visualizing the Power System in Real Time

This attribute would deploy advanced sensors more broadly throughout the system on all critical components. These sensors would be integrated with a real-time communications system through an integrated electric and communications system architecture. The data would need to be managed through a fast simulation and modeling computational ability and presented in a visual form in order for system operators to respond and administer.

Increasing System Capacity

This functionality embodies a generally straight-forward effort to build or reinforce capacity particularly in the high-voltage system. This would include building more transmission circuits, bringing substations and lines up to NERC N-1 (or higher) criteria, making improvements on data infrastructure, upgrading control centers, and updating protection schemes and relays.

Relieving Bottlenecks

This functionality allows the U.S. to eliminate many/most of the bottlenecks that currently limit a truly functional wholesale market and to assure system stability. In addition to increasing capacity as described above, this functionality includes increasing power flow, enhanced voltage support, providing and allowing the operation of the electrical system on a dynamic basis. This functionality would also require technology deployment to manage fault currents.

Enabling a Self-healing System

Once the functionalities discussed above are in place, then it is possible to consider controlling the system in real time. To enable this functionality will require wide-scale deployment of power electronic devices such as power electronic circuit breakers and flexible AC transmission technologies (FACTS). These technologies will then provide for integration with an advanced control architecture to enable a self-healing system.

Enabling (Enhanced) Connectivity to Consumers

The functionalities described above assume the integration of a communications system throughout much of the power system enhancing connectivity to the ultimate consumers. This enhancement creates three new areas of functionality: one which relates directly to electricity services (e.g., added billing information or real-time pricing); one which involves services related to electricity (e.g., home security or appliance monitoring); and the third involves what are more generally thought of as communications services (e.g., data services).

These functionalities will facilitate achievement of the following goals:

- Physical and information assets that are protected from man-made and natural threats, and a power delivery infrastructure that can be quickly restored in the event of attack or a disruption: a "self-healing grid."

- Extremely reliable delivery of the high-quality, "digital-grade" power needed by a growing number of critical electricity end-uses.

- Availability of a wide range of "always-on, price-smart" electricity-related consumer and business services, including low-cost, high-value energy services, that stimulate the economy and offer consumers greater control over energy usage and expenses.

- Minimized environmental and societal impact by improving use of the existing infrastructure; promoting development, implementation, and use of energy efficient equipment and systems; and stimulating the development, implementation, and use of clean distributed energy resources and efficient combined heat and power technologies.

- Improved productivity growth rates, increased economic growth rates, and decreased electricity intensity (ratio of electricity use to gross domestic product, GDP).

A smart grid has the potential to benefit the environment, consumers, utilities and the nation as a whole in numerous ways, as summarized in Figure 1-2. The benefits include the mechanisms for enhanced reliability and power quality as well as energy savings and carbon emission reductions discussed in this book, plus other dividends.

IS THE SMART GRID A "GREEN GRID?"

Today, utilities are struggling to address a new societal and existing or possible future regulatory obligation—mitigating emissions of greenhouse gases, principally carbon dioxide (CO_2), in an effort to curb global climate change and its potentially deleterious implications for mankind.

As part of EPRI's Energy Efficiency Initiative, a first-order quantification of energy savings and carbon-dioxide (CO_2) emissions reduction impacts of a smart grid infrastructure was developed. First-order estimates of energy savings and CO_2 emission reduction impacts were quantified for five applications enabled by a smart grid: (1) continuous commissioning for commercial buildings; (2) distribution voltage control; (3) enhanced demand response and load control; (4) direct feedback on energy usage; and (5) enhanced energy efficiency program measurement and verification capabilities. In addition, first-order estimates of CO_2 emissions reductions impacts were quantified for two mechanisms not tied to energy savings: (1) facilitation of expanded integration of intermittent renewable resources and (2) facilitation of plug-in hybrid electric vehicle (PHEV) market penetration. The emissions reduction impact of a smart grid, based on these seven mechanisms, is estimated as 60 to 211 million metric tons of CO_2 per year in 2030. There are other smart grid improvements which were not part of this analysis, such as a reduction in losses in electricity used in electricity generation and reduced losses by improved area voltage control and other T&D system improvements.

Table 1-1 summarizes the energy savings potentials and corresponding values of avoided CO_2 emissions for each of the seven selected

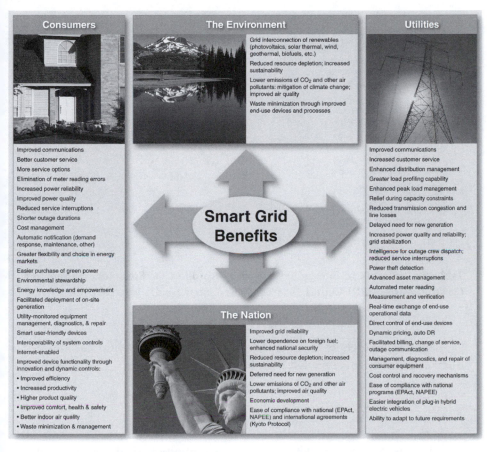

Consumers

Improved communications
Better customer service
More service options
Elimination of meter reading errors
Increased power reliability
Improved power quality
Reduced service interruptions
Shorter outage durations
Cost management
Automatic notification (demand response, maintenance, other)
Greater flexibility and choice in energy markets
Easier purchase of green power
Environmental stewardship
Energy knowledge and empowerment
Facilitated deployment of on-site generation
Utility-monitored equipment management, diagnostics, & repair
Smart user-friendly devices
Interoperability of system controls
Internet-enabled
Improved device functionality through innovation and dynamic controls:
• Improved efficiency
• Increased productivity
• Higher product quality
• Improved comfort, health & safety
• Better indoor air quality
• Waste minimization & management

The Environment

Grid interconnection of renewables (photovoltaics, solar thermal, wind, geothermal, biofuels, etc.)
Reduced resource depletion; increased sustainability
Lower emissions of CO_2 and other air pollutants: mitigation of climate change; improved air quality
Waste minimization through improved end-use devices and processes

Smart Grid Benefits

The Nation

Improved grid reliability
Lower dependence on foreign fuel; enhanced national security
Reduced resource depletion; increased sustainability
Deferred need for new generation
Lower emissions of CO_2 and other air pollutants; improved air quality
Economic development
Ease of compliance with national (EPAct, NAPEE) and international agreements (Kyoto Protocol)

Utilities

Improved communications
Increased customer service
Enhanced distribution management
Greater load profiling capability
Enhanced peak load management
Relief during capacity constraints
Reduced transmission congestion and line losses
Delayed need for new generation
Increased power quality and reliability; grid stabilization
Intelligence for outage crew dispatch; reduced service interruptions
Power theft detection
Advanced asset management
Automated meter reading
Measurement and verification
Real-time exchange of end-use operational data
Direct control of end-use devices
Dynamic pricing, auto DR
Facilitated billing, change of service, outage communication
Management, diagnostics, and repair of consumer equipment
Cost control and recovery mechanisms
Ease of compliance with national programs (EPAct, NAPEE)
Easier integration of plug-in hybrid electric vehicles
Ability to adapt to future requirements

Figure 1-2. Summary of Potential Smart Grid Benefits (Source: EPRI Report 1016905, "The Green Grid," June 2008).

mechanisms in the target year of 2030. Low-end and high-end values are included to show the ranges of savings.

All of the mechanisms combined have the potential to yield energy savings of **56-203 billion kWh** and to reduce annual carbon emissions by **60-211 million metric tons (Tg) CO_2**. On this basis, the environmental value of a U.S. smart grid is equivalent to converting **14 to 50 million** cars into zero-emission vehicles each year.*

---

*Based on an average mid-size sized car driven 12,000 miles per year. Average emissions from Climate Change: Measuring the Carbon Intensity of Sales and Profits." Figure 5: Average CO_2 Emissions Rates by Vehicle Type, 2002). ~ 8,513 lbs CO_2 per car, or ~ 4.25 tons CO_2 per car.

Emissions-Reduction Mechanism Enabled by Smart Grid		Energy Savings, 2030 (billion kWh)		Avoided CO_2 Emissions, 2030 (Tg CO_2)	
		Low	High	Low	High
1	Continuous Commissioning of Large Commercial Buildings	2	9	1	5
2	Reduced Line Losses (Voltage Control)	4	28	2	16
3	Energy Savings Corresponding to Peak Load Management	0	4	0	2
4	Direct Feedback on Energy Usage	40	121	22	68
5	Accelerated Deployment of Energy Efficiency Programs	10	41	6	23
6	Greater Integration of Renewables	--	--	19	37
7	Facilitation of Plug-in Hybrid Electric Vehicles (PHEVs)	--	--	10	60
	Total	56	203	60	211

Table 1-1. Smart Grid Energy Savings and Avoided CO_2 Emissions Summary (2030) ((Source: EPRI Report 1016905, "The Green Grid," June 2008).

Figure 1-3 details the summary of energy-savings and carbon-reduction mechanisms enabled by a smart grid.

ALTERNATIVE VIEWS OF A SMART GRID

It is no surprise that there is no *one* definition of the smart grid. Such a complex machine with so many technology options at hand to improve its functionality is bound to facilitate a variety of broad definitions. As a result, there are a variety of architectures, technologies and configurations already proposed or under formation for what can be described as smart grids. More than 7,000 pilots of some kind are underway today with nearly 1,000 of them over 10 years old. Unfortunately, many of these are not very smart and are limited in scope. Some examples of smart grid visions are as follows:

Capgemini's Vision (www.capgemini.com/energy)
Capgemini believes that in order to make meaningful progress toward addressing the current grid challenges and delivering on the future grid characteristics, utilities should focus on four main activities:

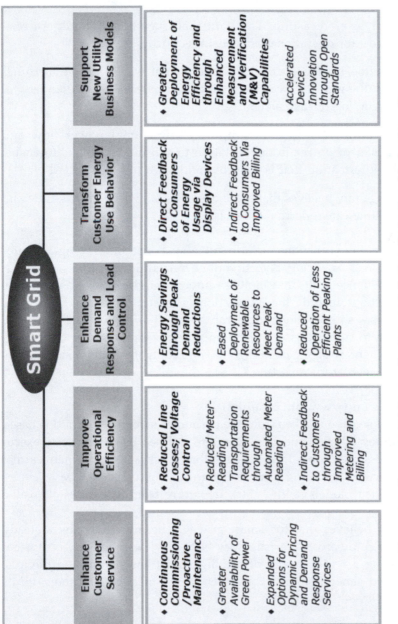

Figure 1-3. Summary of Energy-Savings and Carbon-Reduction Mechanisms Enabled by a Smart Grid (EPRI 1016905, 2008)

1. Gather data: Data should be collected from many sources on the grid.

2. Analysis/forecasting: The data that is gathered should be analyzed—for operational and business purposes.

3. Monitor/manage/act: In the operational world, data that comes from the grid hardware will trigger a predefined process that will inform, log or take action.

4. Rebuilding the grid to support bi-directional power flow and transfer of power from substation to substation: This is to enable the information that is collected and analyzed to be acted on.

Capgemini believes that these activities fall into both real-time and non-real-time categories as follows:

> "Real-time functions include operational and monitoring activities like load balancing, detection of energized downed lines, and high impedance faults and faults in underground cables. Non-real-time functions include the integration of existing and new utility databases so operational data can be fused with financial and other data to support asset utilization maximization and life cycle management, strategic planning, maximization of customer satisfaction, and regulatory reporting."

IBM's Vision (www.ibm.com/iibv)

IBM's vision is taken from the consumer's perspective and is based on a survey of 1,900 energy consumers and nearly 100 industry executives across the globe. IBM believes that it reveals major changes that are underway including a more heterogeneous consumer base, evolving industry models, and a stark departure from a decades-old value chain.

IBM believes that the smart grid will be manifested by a steady progression toward a "Participatory Network," a technology ecosystem comprising a wide variety of intelligent network-connected devices, distributed generation, and consumer energy management tools. IBM's vision believes this includes:

•   Preparing for an environment in which customers are more active participants.

- Capitalizing on new sources of real-time customer and operational information, and deciding which role(s) to play in the industry's evolving value chain.

- Better understanding and serving an increasingly heterogeneous customer base.

To make these improvements, IBM believes that utilities will deploy advanced energy technologies such as smart metering, sensors and distributed generation. They believe that these technologies respond to the following interests:

- The combination of energy price increases and consumers' increased sense of responsibility for the impact of their energy usage on the environment.

- The frequency and extent of blackouts are driving consumers, politicians and regulators alike to demand assessment and upgrade of the industry's aging network infrastructure.

- Climate change concerns have invigorated research and capacity investments in small, clean generating technologies.

- Technology costs have generally decreased as lower-cost communications, more cost-effective computing and open standards have become more prevalent.

IBM sees smart meters, network automation and analytics, and distributed generation driving the most industry changes, from a technological perspective, in the near term.

IntelliGrid^{SM} (www.epri-intelligrid.com)

A consortium was created by EPRI to help the energy industry pave the way to IntelliGrid^{SM}—the architecture of the smart grid of the future. (See Chapter 6 for a more complete description.) Partners are utilities, manufactures and representatives of the public sector. They fund and manage research and development (R&D) dedicated to implementing the concepts of the IntelliGrid^{SM}.

The objective: The convergence of greater consumer choice and rapid advances in the communications, computing and electronics industries is influencing a similar change in the power industry. The

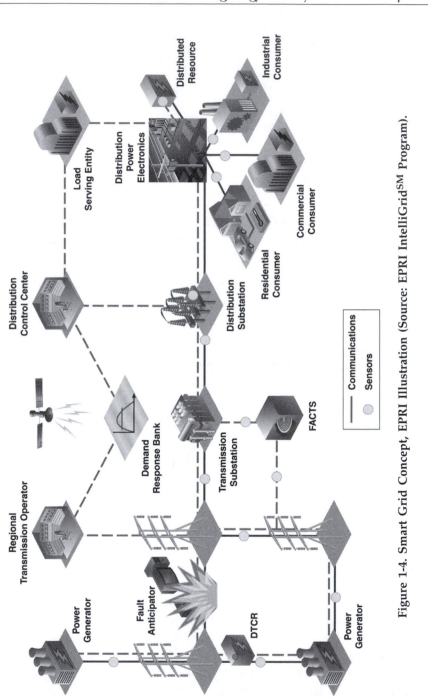

Figure 1-4. Smart Grid Concept, EPRI Illustration (Source: EPRI IntelliGrid^{SM} Program).

growing knowledge-based economy requires a digital power delivery system that links information technology with energy delivery.

The Modern Grid Strategy (www.netl.doe.gov)

The U.S. Department of Energy (DOE) National Energy Technology Laboratory (NETL) is the manager of the Modern Grid Strategy (MGS). The MGS is fostering the development of a common, national vision among grid stakeholders. MGS is working on a framework that enables utilities, vendors, consumers, researchers and other stakeholders to form partnerships and overcome barriers. MGS also supports demonstrations of systems of key technologies that can serve as the foundation for an integrated, modern power grid.

GridWise™ (www.electricdistribution.ctc.com)

The Electric Distribution Program of DOE supports distribution grid modernization, through development and use of advanced sensor, communication, control and information technologies to enable GridWise™ operations of all distribution systems and components for interoperability and seamless integration.

The term GridWise™ denotes the operating principle of a modernized electric infrastructure framework where open but secure system architecture, communication techniques, and associated standards are used throughout the electric grid to provide value and choices to electricity consumers.

DOE's Electric Distribution Program addresses the critical technology area—distributed sensors, intelligence, smart controls, and distributed energy resources—identified in the National Electric Delivery Technologies Roadmap, which defines technology pathways to achieving the Grid 2030 Vision.

DOE's Electric Distribution Program operates the Electric Distribution Transformation (EDT) Program and the GridWise™ Initiative, both within the U.S. Department of Energy Office of Electricity Delivery & Energy Reliability (OE), which is leading a national effort to help modernize and expand America's electric delivery system to ensure economic and national security.

General Electric Vision (www.gepower.com)

General Electric (GE) sees the smart grid "as a family of network control systems and asset-management tools, empowered by sensors, communication pathways and information tools." They envision "A grid that's smarter for all of us" which could experience improvements for utilities, regulators, and businesses.

- For utility executives—GE believes the potential for dramatic energy productivity gains could improve service, control costs and strengthen reliability.

- For operations managers—GE anticipates a reduction in the frequency and impact of outages with improved real-time knowledge of grid status.

- For chief information technology officers—GE sees the smart grid as based on open-standard software and communication protocols, easing systems integration and support.

- For maintenance and engineering professionals- GE believes more can be done with less, and focuses resources on improving service, instead of simply maintaining it. Accurate, real-time and actionable knowledge of grid status enables a shift from time-based to need-based maintenance. It also allows for a more timely response to outages, speeding power restoration.

- For customer service (call center) functions—Calls can be anticipated when an outage has occurred, making systems more responsive to customers. Armed with answers, calls can be resolved faster, allowing delivery of accurate information and a reduction of callbacks, queue times and staffing levels.

GE believes that the smart grid doesn't look all that different from today's grid—on the surface. It's what the smart grid enables that makes the difference. Some major elements GE believes will be included are:

- Distributed generation working seamlessly with current assets.

- Smart homes that make savings practical and ease facets of everyday life.

- Demand response that really knows demand and optimizes response.

- Plug-in vehicles that give back to the grid.

- Upgraded network operations

- Enhanced power quality and security

- Efficiency gains, so you need less and lose less.

Distribution Vision 2010 (DV2010)

DV2010 focuses on distribution reliability improvement. It is a collaboration of six utilities started by WeEnergies to conduct research and development to advance technology for the distribution industry to achieve major reliability improvements.

The DV2010 projects include advanced feeder automation, primary network reconfiguration, accelerated protection and demonstration of a high-reliability network. Several new technologies are being deployed in DV2010 including the use of smart devices to enable enhanced protection, the use of a local automation controller, and a distribution automation system controller incorporating dynamic voltage control, loss reduction and asset optimization.

UK SuperGen Initiative (www.supergen-networks.org.uk)

The consortium faces two broad challenges over two timescales. First, there are engineering problems created by embedding renewable energy sources into a distribution network, and second, there is a need to develop a market and regulatory environment that creates the right commercial drivers to encourage sustainable energy generation and use. These problems need to be solved for the medium-term UK government targets for renewable energy and for the long-term trans-national aspirations for sustainable energy use and climate change. In this context, medium term is 2010 to 2020 and long term is circa 2020 to 2050.
Seven work packages (WP) have been designed to tackle specific issues. Each work package contains a mixture of engineering, economic and social acceptability aspects.

- WP1: System-Wide Reliability and Security
- WP2: Decentralized Operation and control
- WP3: Demand-side Participation
- WP4: Micro-Grids
- WP5: Foresight

- WP6: System Evolution
- WP7: Outreach

Hydro Quebec Automation Initiative

Hydro Quebec's Automation Initiative is based on what they refer to as "advanced distribution system designs " and is intended to be implemented in a structured way (activities, functionalities, technologies) to reduce cost, increase reliability (SAIDI, SAIFI), and increase customer satisfaction. It assumes that some of the gains of the advanced distribution system designs can be measured. Integration of technologies and functionalities is the key to success by reducing costs. Hydro Quebec believes that utilities and customers must have a common vision of advanced distribution system designs through standards to reach cost reduction.

The Galvin Initiative (www.galvinpower.org)

With the inspiration and sponsorship of Robert Galvin (former CEO of Motorola) and his family, this initiative, which began in March 2005, seeks to define and achieve the perfect power system. That is, a consumer-focused electric energy system that never fails. The absolute quality of this system means that it meets, under all conditions, every consumer's expectations for service confidence, convenience and choice will be met. Robert Galvin believes that this will indeed result in the lowest cost system. In the context of this Initiative, the electric energy system includes all elements in the chain of technologies and processes for electricity production, delivery and use across the broadest possible spectrum of industrial, commercial and residential applications.

This initiative identified the most important asset in resolving the growing electricity cost/quality dilemma and its negative reliability, productivity and value implications as technology-based innovation that disrupts the status-quo. These innovation opportunities begin with the consumer. They include the seamless convergence of electricity and telecommunications service; power electronics that fundamentally increase reliability, controllability and functionality; and high-power quality microgrids that utilize distributed generation, combined heat and power (CHP) and renewable energy as essential assets. This smart modernization of electric energy supply and service would directly empower consumers and, in so doing, revitalize the electricity enterprise for the 21st century. This focus on the consumer interface also reflects the relatively intractable nature of the highly regulated bulk power infrastructure that now dominates U.S. electric energy supply and service.

This initiative defined a set of four generic electric energy system configurations that have the potential to achieve perfection, together with the corresponding innovation opportunities that are essential to their success. Each candidate configuration reflects a distinct level of electric energy system independence/interconnection from the consumer perspective. In so doing, these configurations address, in a phased manner, the fundamental limitations to quality perfection in today's centralized and highly integrated U.S. electric power system. They also provide a variety of new avenues for engagement by entrepreneurial business innovators.

The four generic system configurations identified by this initiative are (1) device-level (portable) systems, which serve a highly mobile digital society; (2) building integrated systems, which focus on modular facilities serving individual consumer premises and end-use devices; (3) distributed systems, which extend the focus to local independent microgrids, including interconnection with local power distribution systems; and (4) integrated centralized systems, which fully engage the nation's bulk power supply network. These configurations should be considered as a complementary series rather than as competing systems.

Refer to Chapter 4 for a more complete description of the Galvin Electricity Initiative.

Electricite de France (EDF) Power-Strada

EDF proposes to "invent the smart grid." It defines it as integrating distributed energy resources with dispersed intelligence and advanced automation. To accomplish this objective, it has identified a series of technology challenges intended to:

- Build a vision for the long-term research program for the EDF group.
- Help the decision-making process in providing strategic highlights.
- Offer a better visibility and understanding of their activities.
- Maintain the right balance between short- and long-term activities.

European Union Smart Grid (www.smartgrids.eu)

The European Union has initiated a smart grid effort. Their vision is:

- To overcome the limits on the development of distributed generation and storage.
- To ensure interoperability and security of supply.
- To provide accessibility for all the users to a liberalized market.
- To face the increased complexity of power system operation.
- To reduce the impact of environmental consequences of electricity production and delivery.
- To enable demand-side participation.
- To engage consumers interest.

The European Union is undertaking various activities to overcome barriers to the development of European smart grids. They believe that a clear and stable regulatory framework stimulating the development of smart grids is needed. This must be coupled with the development of a common market model and conditions to be applied to the electricity sector throughout the EU by national governments and regulators. European standardization bodies must define common technical requirements supporting smart grids.

The EU should provide directives, members states and national regulators need to define the right incentives scheme applied both to network operators and consumers to promote the implementation of smart grids utilizing available technology and late-stage R&D products and applications.

The European Union has initiated a cluster of seven research projects centered primarily on enabling the integration of distributed generation (especially renewables) into the European power system. Several of them are described in the following sections.

Dispower Project (www.dispower.org)
        The Dispower Project is an elaboration of strategies and concepts for grid stability and system control in distributed generation (DG) networks. It includes preparation of safety and quality standards in DG networks and investigations on quality improvements and requirements by decentralized inverters and generation systems. Assessments of impacts to consumers by s, energy trading and load management are conducted including development of planning and design tools to ensure reliable and cost-effective integration of DG components in regional and local grids and creation of Internet-based information systems for improved communication, energy management and trading. Investigations on

contract and tariff issues regarding energy trading and wheeling, and ancillary services are part of the analysis. Objectives include improvement and adaptation of test facilities and performance of experiments for further development of DG components, control systems, as well as design and planning tools, and successful dissemination and implementation of concepts and components for an improved integration of DG technologies in different European electrical network environments.

MicroGrids Project (www.microgrids.power.ece.ntua.gr)

The MicroGrids Project is a study of the design and operation of microgrids so that increased penetration of renewable energies and other micro sources including fuel cells, micro-turbines and CHP emissions will be achieved and CO_2 emissions reduced. It includes development and demonstration of control strategies that will ensure the operation and management of microgrids is able to meet the customer requirements and technical constraints (regarding voltage and frequency) and is delivering power in the most efficient, reliable and economic way. This involves determination of the economic and environmental benefits of the microgrid operation and proposal of systematic methods and tools to quantify these benefits and to propose appropriate regulatory measures.

The definition of appropriate protection and grounding policies that will assure safety of operation and capability of fault detection, isolation and islanded operation are part of the work scope, as well as identification of the needs and development of telecommunication infrastructures and communication protocols required to operate such systems (this investigation will include consideration of the possibility of exploiting the power wires as physical links for communication purposes). The project managers intend to simulate and demonstrate microgrid operation on laboratory models. Finally, they will deploy microgrids on actual distribution feeders operating in Greece, Portugal and France quantifying via simulation the environmental, reliability and economic benefits from their operation.

CONCLUSION

While no one definition of the smart grid prevails, it can be concluded that the smart grid may more appropriately be named the "smarter grid." A variety of approaches are presented in this chapter—all of which have a common theme of improving the overall functionality of the power delivery system. Some of these improvements advocate incremental change, some only focus on one element of automation. But all collectively envision a system which improves the environment, enhances the value of electricity, and improves the quality of life.

References
The Green Grid, EPRI Report 1016905, June 2008.

Plugging in the Consumer: Innovating Utility Business Models for the Future, Publication G510-7872-00,© IBM 2007.

Chapter 2

Electric Energy Efficiency in Power Production & Delivery

INTRODUCTION

As interest in electric energy efficiency increases, practitioners and others interested in energy efficiency are questioning whether the scope of end-use energy efficiency should be broadened to electric uses in power delivery systems and in power plants. Interestingly, these two aspects of efficiency are only briefly mentioned in any of the smart grid initiatives currently in play. Efficiency advocates are questioning whether investments made within power delivery systems and power plants to reduce electricity demand may be as advantageous as those made in end-use efforts with consumers. Power plant improvements in electricity consumption could include methods to improve overall efficiency, therefore, increasing the heat rate (decreasing Btu input per kWh output) of the power plant. Power delivery system improvements could include use of efficient transformers and better voltage and reactive power control, among others. Each of these are directly related to an informed view of the smart grid.

One element of the smart grid that relates to the efficient production of electricity has to do with condition monitoring and assessment. In condition monitoring and assessment, sensors and communications are used to monitor plant performance and to correlate that performance to historic data, theoretical models and comparable plant performance. The objectives are to optimize performance and manage maintenance. Most plants have some form of condition monitoring. A number of power plant operators subscribe to commercial data services which enable comparisons of individual plant performers to a central database. A few plant operators have centralized data collection and monitoring. The concept of the expanded use of sensor, communications and computational ability is part of the smart grid concept. As such, a truly smart grid will include extending this concept all the way from power production, through delivery to end use.

This chapter focuses on the unique aspects of power production and delivery related to efficiency. Again, these are largely enabled by the smart grid.

POWER PLANT ELECTRICITY USE

Electricity is used in power plants in several ways. The principal use is in powering motors which drive pumps, fans and conveyors. Other uses include information technology (computers, communications and office equipment), building lighting, heating, domestic water heating, ventilation, air conditioning, food service, environmental treatment of waste streams, and select use of electric immersion heating to augment power production.

Each of these provide opportunities to apply cost-effective electric energy-efficient technologies to reduce overall electricity use in power production. However, the dominance of electricity use is in motors.

Data are sketchy and limited, but a few studies have indicated that typically 5 to 7% of electric energy produced in steam power plants (coal, biomass, gas and nuclear) is used on-site to enable electricity generation. Less is used in combustion turbine power plants, and even less in hydro power and wind. Much of that electricity use is in electric motor-driven fans and pumps, but there are a number of other uses ranging from lighting to space conditioning and digital devices.

Pumping and fan applications consume a large portion of on-site energy and are, therefore, a key opportunity. Both pumps and fans must be regulated with throttling valves and dampers to respond to generator loading and climatic conditions. Both pumps and fans are well suited for adjustable speed drives (ASDs) to regulate air and fluid flow and to operate at optimum efficiency. Running pumps and fans with adjustable speed drives yields substantial energy savings over regulating fluid or air flow with valves and dampers. Again, the smart grid concepts are the catalyzing element. Controlling motor drives and monitoring their condition requires sensors, communications, and control.

When less energy is consumed within the plant, generating capacity is released to sell to consumers without changing the rating of the generator, turbine or boiler. This reduces total energy requirements and corresponding CO_2 and other emissions. In addition, the cost of power generated is reduced.

Major stationary sources of air pollution and major modifications to major stationary sources are required by the Clean Air Act to obtain an air pollution permit before commencing construction. The process is called new source review (NSR) and is required whether the major source or modification is planned for an area where the national ambient air quality standards (NAAQS) are exceeded (nonattainment areas) or an area where air quality is acceptable (attainment and unclassifiable areas).

As a result, implementing various actions to reduce on-site electricity consumption in power plants is "included" in NSR applications in new plant construction or where plant upgrades are planned. In existing plants where these actions may be considered, the point of contention is around whether these efficiency modifications are considered to be "major" modifications. At present, this is not well defined. On occasion, regulatory agencies have considered even "minor" changes in plant configurations to be major modifications. As such, plant owners are reluctant to open the entire plant operations to scrutiny for the sake of the on-site electricity savings. The result is that improvements which would have resulted in efficiency improvements and emissions improvements are not aggressively pursued.

POWER PLANT LIGHTING

Lighting has a high potential for technology improvement to enhance its functionality in power plants. Offices, laboratories and control rooms in power plants largely tend to incorporate higher-efficiency fluorescent lighting, while the boiler house and turbine halls often employ high-pressure sodium, metal halide and other high-intensity discharge lighting. However, many older plants still use mercury or older metal halide lighting systems. New metal halide systems could substantially improve lighting and reduce energy requirements. Light-emitting diodes (LEDs) are emerging in many places such as in exit signs and indicator panels. Potential areas for lighting improvement include energy efficiency, light quality, aesthetics, automation, longevity, convenience and form factor.

Lighting system efficiency is a function of several factors. The overall efficiency of the system depends on the efficacy* of the light source (measured in lumens of light output per watt of input power), the effectiveness of the fixture in delivering the generated light to the area it is desired, and the ability of the controls (if any) to adjust the light level as parameters such as occupancy, day lighting, security and personal preference dictate. Efficiency is maximized as the performance of each of these factors is optimized. Again, the integration of controls is an extension of the smart grid.

Much of the artificial lighting in place today is considerably less efficient than theoretically, or even practically, possible. Conventional incandescent lamps are a good example of this—roughly 95% of the electricity entering an incandescent lamp is rejected as heat. Typical incandescent bulbs produce only 10 to 23 lumens per watt. Conventional fluorescent lighting systems offer a substantial efficiency improvement over incandescent lamps, converting 20 to 30% of electricity into light; however, 70 to 80% of the input energy is still rejected as heat. Historically, considerable effort has been focused on improving lighting efficiency, and research continues.

One example of a significant increase in efficiency in power plant offices is illustrated in the transition from T12 (1.5-inch diameter) fluorescent lamps with magnetic ballasts to T8 (1-inch diameter) fluorescent lamps with electronic ballasts. This transition began to occur in the late 1970s and early 1980s. Now T8 or even newer T5 electronic ballast systems are the standard for new construction and retrofits. The efficiency improvement, depending on the fixture is roughly 20 to 40% or even more. For example, a two-lamp F34T12 fixture (i.e., a fixture with two 1.5-inch diameter, 34 W lamps) with standard magnetic ballast requires 82 W, while a two-lamp F32T8 fixture (i.e., a fixture with two 1-inch diameter, 32 W lamps) with electronic ballast requires only 59 W, which is an electricity savings of 28%. The savings is attributable to the lower wattage lamps as well as the considerably more efficient ballast. Next generation T5 (5/8-inch diameter) fluorescent lamps continue to improve lighting efficiency and can replace conventional lamps in certain applications, such as indirect lighting.

---

*Lighting efficacy is the ratio of light output of the lamp in lumens to the input power in watts. In the lighting industry, efficacy is a more meaningful measure of the lamp output than efficiency (which is the ratio of the useful energy delivered to the energy input).

Increased penetration of high-efficiency lighting systems that combine efficacious light sources with fixtures that effectively direct light where it is desired, coupled with automated controls to dim or turn the lighting on and off as needed, will collectively act to improve overall lighting efficiency.

Maintenance Issues

Developers are continually striving to improve the longevity of lamps. For example, a typical fluorescent lamp's lifetime is 8,000 hours, while LEDs recently launched have a useful life* of 50,000 hours or more—a six-fold increase in comparison to the fluorescent lamp. Future lighting technologies may be able to achieve useful life expectancies of four million hours. Furthermore, with greater life expectancies in lamps, there will likely be an increase in the number of applications employing lighting. Previously, the inconvenience of frequent lamp changes precluded lighting in certain applications. However, the benefit of illumination in newer applications (e.g., monitoring and indicator lighting, improved outside plant area lighting, etc.), coupled with lover energy requirements, may start to outweigh the maintenance and cost issues that were formerly a barrier.

Lamps also suffer from creeping old age—the longer the lamp burns, the fewer lumens per watt it produces. Lamp lumen depreciation is a partial factor in determining the maintenance (light loss) factor, which is an important step in lighting calculation and design. The change in lumen output is often referred to as lumen maintenance. Newer technologies improve the lumen maintenance factor.

Lighting technologies of particular interest to applications in power plants must exhibit improvements in one or more of the attributes required to meet evolving plant operator needs and expectations (e.g., efficiency, light quality, aesthetics, automation, longevity, form factor, convenience, etc.). Table 2-1 lists the lighting technologies which may be considered for application in power plants so as to reduce energy consumption.

---

*Useful life refers to the life the lamp operates at an acceptable light level. The illumination output from LEDs slowly diminishes over time. The end of an LED's useful life occurs when it no longer provides adequate illumination for the task; however, the LED will continue to operate, albeit at a reduced light level, for an indefinite amount of time.

Table 2-1. List of Lighting Technologies Applicable for Power Plant Energy Efficiency Improvement

Induction Lamps	Multi-photon Emitting Phosphor Fluorescent Lighting
T-5 Fluorescent Lamps	Metal Halide High-intensity Discharge Lamps
High-Pressure Sodium High-intensity Discharge Lamps	

POWER PLANT SPACE CONDITIONING AND DOMESTIC WATER HEATING

Space conditioning and domestic water heating are other two technology application areas which use considerable electricity in power plants. Workers in offices and laboratories require space conditioning to create comfortable environments in which to work. When properly designed, space-conditioning systems afford the worker a healthy environment that enables productivity and a sense of well-being. Spaces that are either too hot or too cold make it more difficult for individuals to carry out tasks. Such environments can also lead to adverse health effects in occupants. Domestic water heating is also essential for the comfort and well-being of workers. Hot water is used for a variety of daily functions, including bathing, laundry, food preparation and in laboratories.

Electricity is used in a variety of ways in space conditioning and domestic water heating (Smith, et al. 2008). In space conditioning, the primary function is to heat, cool, dehumidify, humidify and provide air mixing and ventilating. To this end, electricity drives devices such as fans, air conditioners, chillers, cooling towers, pumps, humidifiers, dehumidifiers, resistance heaters, heat pumps and electric boilers. Electricity also powers the various controls used to operate space-conditioning equipment. In water heating, electricity runs electric resistance water heaters, heat pump water heaters, pumps and emerging devices such as microwave water heaters.

Power plant space-conditioning systems may implement packaged roof-top or ground-mounted units, or a central plant. The majority of space-conditioning electricity use in power plants is attributed to cooling equipment rather than heating equipment. The most common space conditioning systems for power plants are of the unitary type, either single-package systems or split systems. These are used for cooling approximately two-thirds of the air-conditioned spaces in U.S. power plants. For very large office buildings associated with power plants, direct-expansion systems, absorption chillers or central chiller plants are used.

Conventional electric water heating systems generally have one or two immersion heaters, each rated at 2 to 6 kW depending on tank size. These systems store hot water in a tank until it is needed. Newer systems generate hot water on demand.

Great strides have been made in the last few decades to improve the efficiency of space conditioning and water heating equipment. The most notable advancements have been in space cooling equipment—largely in vapor compression cooling. Improvements in the efficiency of space cooling systems have a large impact on electricity use in power plants with significant office and laboratory space, since they account for the majority of electricity use in these spaces. Much of the progress in space cooling efficiency is due to federal standards that dictate the minimum efficiency of new air-conditioning systems. Though the volume of space that is mechanically cooled is on the rise, corresponding increases in electricity consumption have been tempered by efficiency improvements in air-conditioning systems and improvement in building thermal integrity.

Larger commercial-scale air-conditioning systems are often rated with energy-efficiency ratio (EER) and integrated part-load value (IPLV) parameters. Table 2-2 lists the minimum federal standards for larger-scale units (>65,000 Btu per hour) as of October 29, 2001. Manufacturers sell systems with a broad range of efficiencies. For example, in the 65,000 to 135,000 Btu per hour capacity range, it is possible to buy units with an EER as high as 12.5, even though the current federal standard is 10.3 (Smith, et al., 2008). Units with high EERs are typically more expensive, as the greater efficiency is achieved with larger heat exchange surface, more efficient motors, and so on.

One of the most efficient space-conditioning devices is the heat pump. Heat pumps can also be used for water heating and in integrated systems that combine water heating and space conditioning. The heating performance of a heat pump is measured by the heating seasonal performance factor (HSPF), which is equal to the number Btus of heat added per watt-hour of electricity input. HSPF values for commercially available heat pumps are 6.8 to 9.0 and higher for the most efficient

Table 2-2. Standards for Commercial-scale Air-conditioning Systems, 2001

Equipment Size (Btu/hour)	EER	IPLV
65,000 to <135,000	10.3	
135,000 to <240,000	9.7	
240,000 to <760,000	9.5	9.7
760,000 and larger	9.2	9.4

systems. The cooling performance of heat pumps is measured with the SEER value (as is used for air conditioners). Values of 10.0 to 14.5 and higher are typical for the most efficient systems. New federal standards took effect in 2006 which raised the HSPF for heat pumps from the 1992 minimum value of 6.8 to a new minimum value of 7.8. In addition, the minimum SEER values have been increased from 10.0 to 13.0 (same as for air conditioners). Many existing older units have SEERs of 6 to 7, or roughly half the new minimum requirement. Therefore, substantial efficiency improvements are possible by replacing older equipment.

Chillers are used in central space-conditioning systems to generate cooling and then typically distribute the cooling with chilled water to air-handling or fan-coil units. Chillers are essentially packaged vapor compression systems. The vapor compression refrigeration cycle expands, boils, compresses, and re-condenses the refrigerant. The refrigeration system is closed, and the refrigerant flows between high-pressure, high-temperature vapor and low-pressure, low-temperature liquid. First, the liquid refrigerant flows from the condenser to the expansion valve, which reduces the liquid pressure before it enters the evaporator. Second, the evaporator coils absorb heat from the building's chilled water loop that runs through the evaporator. This heat causes the liquid to expand and become a vapor. The refrigerant vapor then leaves the evaporator and flows to the compressor. The compressor maintains a pressure difference between the evaporator and condenser. In the compressor, the pressure and the temperature of the vapor increase before the vapor can be condensed by relatively warm water or air. Finally, in the condenser, the vapor releases the heat it absorbed in the evaporator and the heat added by the compressor, and it becomes a liquid again. This is accomplished either via a water-cooled or an air-cooled condenser.

Chillers are often the largest single user of electricity in a large office space-conditioning system. Therefore, improvements in chiller efficiency can have a large impact of electricity use. Chiller efficiency is specified in units of kW per ton of cooling. It is normally quoted based on the loading application—either full-load or part-load. In full-load applications, the load on the chiller is high and relatively constant (e.g., baseline chillers), and the full-load efficiency is used to measure performance. For part-load applications, which are more common, the load on the chiller is variable, and use of the part-load efficiency is a more meaningful measure of performance. The choice of both the compressor and the condenser affects the overall efficiency of the chiller. Some chillers use air-cooled condensers, but most large units operate with evaporative cooling towers. Cooling towers have the advantage of rejecting heat to a lower temperature heat sink because the water approaches the ambient wet-bulb temperature while air-cooled units are limited to the dry-bulb temperature. As a result, air-cooled chillers have a higher condensing temperature, which lowers the efficiency of the chiller. In full-load applications, air-cooled chillers require about 1 to 1.3 kW or more per ton of cooling, while water-cooled chillers usually require between 0.4 to 0.9 kW per ton. Air-cooled condensers are sometimes used because they require much less maintenance than cooling towers and have lower installation costs. They can also be desirable in areas of the country where water is scarce and/or water and water treatment costs are high since they do not depend on water for cooling.

The efficiency of water heaters is measured with a quantity called the energy factor (EF). Higher EF values equate to more efficient water heaters. Typical EF values range from about 0.7 to 9.5 for electric resistance heaters, 0.5 to 0.8 for natural gas units, 0.7 to 0.85 for oil units, and 1.5 to 2.0 for heat pump water heaters (Smith, et al., 2008). Since there are no flame or stack losses in electric units, the major factor affecting the efficiency of electric water heaters is the standby loss incurred through the tank walls and from piping. The heat loss is proportional to the temperature difference between the tank and its surroundings. Newer systems produce hot water on demand, eliminating the storage tank and its associated losses.

Building Infiltration

A major cause of energy loss in space conditioning is due to air entering or leaving a conditioned space (Global Energy Partners, 2005).

Unintentional air transfer toward the inside is referred to as *infiltration,* and unintentional air transfer toward the outside is referred to as *exfiltration.* However, *infiltration* is often used to imply air leakage both into and out of a conditioned space, and this is the terminology used here. In a poorly "sealed" building, infiltration of cold or hot air will increase heating or cooling energy use. Air can infiltrate through numerous cracks and spaces created during building construction, such as those associated with electrical outlets, pipes, ducts, windows, doors and gaps between ceilings, walls, floors and so on. Infiltration results from temperature and pressure differences between the inside and outside of a conditioned space caused by wind, natural, convection and other forces. Major sources of air leakage are plenum bypasses (paths within walls that connect conditioned spaces with unconditioned spaces above, or adjacent to the conditioned space), leaky duct work, window and door frames, and holes drilled in framing members for plumbing, electrical and HVAC equipment.

To combat infiltration and reduce energy losses from heating and cooling systems, wraps, caulking, foam insulation, tapes and other seals can be used. Sealing ducts in the building is also important to prevent the escape of heated or cooled air. Caution must be exercised to provide adequate ventilation, however. Tight spaces often require mechanical ventilation to ensure good air quality. Standards vary, depending on the type of occupancy.

Complaints about the temperature from occupants of office spaces in plants are all too common. Studies show employees find that one-third to one-half of office buildings are too cold or too warm (Kempton, et al., 1992). Occupants find themselves dressing for winter during the summer to prevent being too cold from an over-active cooling system, or taking off layers of clothing during the winter because the heating system seems to be running "full blast." The temperatures are also variable throughout a given building. There are often hot spots and cold spots. In addition, humidity is typically not controlled effectively, nor is the level of indoor air contaminants.

Environmental pressures are changing the form of space conditioning and water heating devices. For example, traditional space-cooling technologies rely on vapor compression cycles that use ozone-depleting refrigerants, such as chlorofluorocarbons (CFCs) and hydrochlorofluorocarbons (HCFCs). These refrigerants are being phased out. (1995 was the last year than CFC refrigerants could be legally manufactured in most of the industrialized world. HCFC refrigerants will be produced until 2030.) The Kyoto Protocol is driving the adoption of environmentally friendly space-cooling technologies in several countries around the world, most notably in Japan and Scandinavia. In the future, space-cooling systems will increasingly rely on environmentally friendly "natural" refrigerants (one possible alternative is CO_2) or they may use entirely new space-cooling technologies (e.g., magnetic refrigeration and thermo-tunneling).

Space-heating and water-heating systems that rely on the combustion of fossil fuels are also being replaced by cleaner alternatives. Electricity is an efficient source of energy that is very clean at the point of use. It can be used to drive heat pumps for space conditioning and for domestic water heating. It can also be applied to novel heating devices such as microwave water heaters.

Table 2-3 lists the technologies that should be considered when assessing electric energy improvements for conditioned spaces in power plants.

MOTORS

Processes driven by electric motors typically consume 80% of the electricity used in electricity production. Implementing electric adjustable speed drives (ASDs) on motors in power plants improves the heat rate by increasing the efficiency of these processes. ASDs increase plant availability and flexibility through improved process control and reduce emissions and maintenance costs. ASDs can be used in power plants for boiler feedwater, cooling water, circulation water pumps as well as forced draft (FD) and induced draft (ID) fans. In gas turbine power plants, they can also be used for gas turbine (GT) starters, drives for gas booster compressors, boiler heat recovery steam generators (HRSG) feedwater pumps and cooling water pumps.

As the demand for electrical energy varies, the required flow of a fan or a pump varies due to changes in ambient conditions or fuel properties. Due to these varying conditions, a continuous control of the processes and equipment such as centrifugal fans and pumps, is required.

Processes driven by pumps or fans are usually controlled mechanically with inlet guide vanes, throttling valves or hydraulic couplings.

Space Conditioning	
Electrically Heated Windows	Series Desiccant Wheel for Improved Dehumidification
Electrically Heated Floors	Hydronic Dry Floors
Residential Two-Stage, Condensing Gas Furnaces	Condenser Heat Reactivated Desiccant for Improved Dehumidification
Smart Thermostats	Thermotunneling-Based Cooling
Active Magnetic Regenerative (AMR) Cooling	Ultraviolet Germicidal Irradiation (UVGTI) of Chiller Coils
Stirling Engines	Demand-Controlled Hybrid Ventilation
	Microwave Water Heating and Purification
Domestic Water Heating	**Domestic Water Heating and Space Conditioning**
Concentration Solar	Heat Pump Water Heating/Space Conditioning
Space Conditioning and/or Water Heating Using Carbon Dioxide (CO_2) Refrigeration Cycle	

Table 2-3. List of Space Conditioning and Domestic Water Heating Technologies

The motors are driven at a fixed speed making it practically impossible to achieve the optimal process efficiency over a control range. With electric adjustable speed drives, changing the mechanical output of the system is achieved by changing the motor speed. This saves energy, decreases CO_2 and other emissions from electricity production and minimizes operating costs.

Since pumps and fans typically run at partial mechanical load, substantial energy savings can be achieved by controlling motor speed with adjustable speed drives. A small reduction in speed correlates to a large reduction in the energy consumption. A pump or a fan running at half speed consumes one-eighth of the energy compared to one running at full speed. For example, according to ABB (www.ABB.com), by employing adjustable speed drives (inverter technologies) on centrifugal pumps and fans instead of throttling or using inlet guide vanes, the energy bill can be reduced by as much as 60%.

Although a number of ASD technologies have been found to exist commercially for large motors, three U.S. designed technologies have survived the rigors of competition. They are as follows:

- Current-source inverter
- Modified load commutated inverter (modified LCI)
- Current-source gate turn-off thyristor-pulse-width modulated (GTO-PWM)

Of these three, the latter two are still being successfully sold for large motors (2000 HP and larger). The current-source system has been shown to have a number of excellent features. It has good harmonic control when used in a 12-pulse input, 12-pulse output configuration, it has no output filter capacitor requirement, and can be used in a full regenerative braking mode.

The modified LCI inverter, another form of current-source technology, has provided an economic ASD system that has shown the power production industry that it can reduce fuel costs in many motor applications with electronic speed control. The modified LCI system has the simplicity afforded by a rectifier and inverter using the same components. The DC link diverter circuit provides for inverter commutation when the output filter capacitor can no longer provide excitation to the induction motor to allow LCI operations. This system has been packaged with water cooling of the power electronics to simplify the overall cooling system. Water cooling is important for many power plant applications where the air is contaminated with coal or ash dust.

EPRI DEMONSTRATIONS

EPRI has conducted research programs to study the application of high-power adjustable speed drives (ASDs) to auxiliary motors of electric generating stations (EPRI CU-9614, 1990). Four utilities participated in field tests of these large ASDs on boiler feed pumps and forced draft fans.

In the period of the field tests, rapid advances had been made in the technology of ASDs for large induction motors. The principal gains arose from new schemes for commutating the inverter. Vector control has also been introduced to allow separation of motor flux control and motor current control to allow the control of torque separately from voltage. These field test projects use the existing power plant squirrel cage induction motors. The power electronics equipment additions allowed control of feedwater flow or air flow directly by motor speed, thus eliminating the control valve or inlet vanes and the power losses associated with these devices.

The test program observed the performance of the equipment operating with and without ASD control. Power measurements were made to verify power savings and economics. Harmonics were measured at the input and output of the ASD. Motor vibrations were measured over the speed range. Current and voltage wave shapes were recorded and means were established to determine ASD efficiency. Several ASD cooling systems and enclosures were evaluated in the course of the test program.

These tests occurred over a five-year period. During this time, modified load-commutated inverters and the current-source GTO-PWM inverters were installed in over 200 installations nationwide, ranging from 600 HP to 9000 HP. GTO stands for gate turn-off thyristors; PWM stands for pulse-width modulated—a technique for creating low harmonic content AC waveforms.

These field tests yielded a wealth on the application of this technology to large power plant induction motors. Among the lessons learned in this work were the following:

- The potential for operating cost savings by controlling process flow with motor speed is real.

- Reliability of large ASDs is not an issue. Several improvements in ASDs that have developed directly from operating experience have contributed to improved overall system reliability:
  — An input transformer is now used on all large ASD installations to control common-mode voltage.
  — Shaft torsional resonance caused by interaction of the ASD output capacitor filter and the motor winding is now understood and can be controlled either by eliminating harmful output harmonics with the GTO-PWM ASD or a 12-pulse inverter or by separating the electrical and mechanical resonance frequencies with an output reactor.
  — Power electronic devices, like thyristors and GTOs, have proven to be robust and reliable when correctly applied.
  — Available control systems have proven to provide trouble-free service.

Utilizing the information gained from these tests, a preferred configuration for a power plant-specific ASD has evolved. This configuration has the following features:

- Input transformer to control input harmonics and line-to-ground voltage at the motor.

- Uninterruptible power supply (UPS) system for clean power to the ASD control system to eliminate adverse effects from voltage spikes, dips, and interruptions on ASD performance.

- Water-cooled thyristors to simplify cooling of the ASD in the often hot, dusty environment of power plants.

- GTO thyristor inverter to control harmonics to the motor in order to eliminate shaft torsional resonance.

- Ground between inverter and motor to control line-to-ground voltage at the motor used in combination with input transformer.

The preferred configuration for a large power plant induction motor ASD is shown in Figure 2-1.

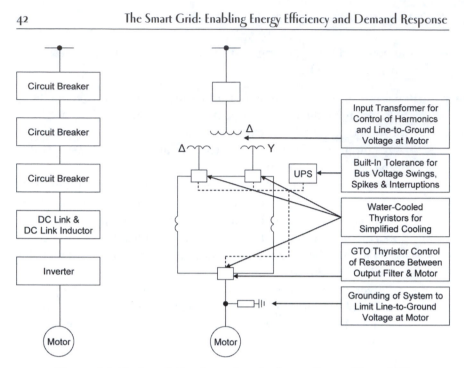

Figure 2-1. Preferred Configuration for Large Power Plant ASD

There have been remarkable advances in power electronics control of large induction motors in power plant applications. Different housings, different cooling systems, different rectifier arrangements, and different inverter technologies have been analyzed, and preferred ASD application procedures have been identified.

However, the application of adjustable speed drive to large induction motors is still a relatively new technology.

Utility-specific ASD

The lessons learned in EPRI field test installations and from several other recent utility installations of induction motor ASDs have contributed to a concept for a second-generation ASD specifically for power plants. Features of the power plant-specific ASD are as follows:

•   Use of input transformer for 12-pulse or 18-pulse converter.

•   Grounding of ASD system to stabilize DC link voltage and eliminate motor over-voltage.

- Built-in design tolerance for bus voltage swings, spikes and inter-ruptions.

- Water-cooled thyristors for simplified and more effective cooling.

- Control of or elimination of resonance between output filter and motor.

EFFICIENCY IN POWER DELIVERY

The improvement in the power delivery system (electric transmission and distribution) through the use of smart grid technologies can provide significant opportunities to improve energy efficiency in the electric power value chain. The transmission and distribution system is responsible for 7% of electricity losses. On the distribution system, data from many utilities clearly prove that tighter voltage control reduces energy consumption. As a result, a number of utilities have applied the concept of conservation voltage reduction (CVR) to control distribution voltage and reduce end-use energy consumption without adverse effects. At the same time, the U.S. Department of Energy (DOE) recently issued a final rule (October 2007) that mandates efficiency standards for various types of distribution transformers beginning in 2010. DOE estimates that these standards will reduce transformer losses and operating costs by 9 to 26%. On the customer side of the meter, the opportunities for improving energy efficiency are many and varied, across all sectors of the economy.

CONSERVATION VOLTAGE REDUCTION

On the distribution system, evidence suggests that utilities may be able to achieve dramatic energy and demand savings by operating distribution feeders at the lower end of acceptable service ranges through smart grid applications. This concept is called conservation voltage reduction (CVR). Despite considerable utility research on this subject in the 1970s and 1980s, few recent studies have examined this potential and the means to attain it. By some estimates, only 7.5% of U.S. feeders incorporate the necessary technologies to tightly control voltage in this fashion.

For a typical U.S.-based, 120-V nominal service voltage (as measured at the customer meter), the governing U.S. standard is the American National Standards Institute (ANSI) standard C84.1, which specifies a preferred range of +/- 5%, or 114-126 V. Utilities tend to maintain the average voltage above 120 V to provide a larger safety margin during periods of unusually high loads, as well as to maximize revenue from electricity sales. Utilities generally regard 114 V as the lowest acceptable service voltage to customers under normal conditions, since a 4-volt decrease is typically assumed from the customer meter to the plug, and most appliances are designed to operate at no less than 110 V of delivered voltage.

In general, lowering voltage decreases load, and thereby saves energy. "CVR Factor," which is defined as the percentage reduction in power resulting from a 1% reduction in voltage, is the metric most often used to gauge the effectiveness of voltage reduction as a load reduction or energy savings tool. The CVR factor differs from utility to utility and circuit to circuit based on each circuit's unique load characteristics. Empirical data from utilities across the United States suggests that CVR factors range from 0.4 to 1.0, and in some cases, may even slightly exceed 1.0. EPRI recently completed a calculation of the mean CVR factor (0.8), based on equal weighting of 13 utility CVR factors. One key characteristic that determines the effectiveness of voltage regulation for load reduction is the amount of resistive versus reactive load in a given circuit.

Distribution system efficiency (DSE) refers to a range of electric utility measures designed to modify the voltage delivered to end-use customers to a range lower than or tighter than the ANSI standard C84.1.

Resistive loads, such as electric resistance space and water heaters and incandescent lamps, act as resistors and predominantly draw real power. As a result, resistive loads respond directly with voltage changes—lower voltages result in reduced power consumption. In fact, power use in an individual resistive load is proportional to the square of the voltage, meaning that the CVR factor will be greater than 1.0 as long as the load is on. However, automatic controls on resistive loads such as space and water heaters usually reduce this impact, in aggregate, over a large number of loads by keeping heater elements "on" for longer periods to maintain temperatures. Despite this phenomenon, power use still varies directly with the voltage for resistive loads.

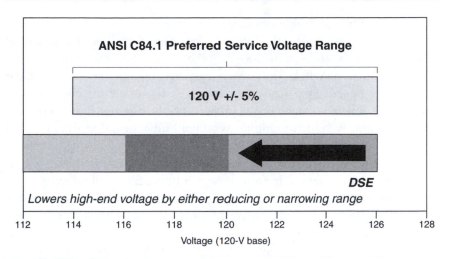

Figure 2-2. Distribution System Efficiency (DSE) in the Context of the ANSI C84.1 Preferred Service Voltage Standard for 120-V Systems. (Global Energy Partners, 2005)

Reactive loads, typically containing significant inductance (also known as inductive loads), include motors, pumps, and compressors, which draw both real and inductive-reactive power. Reducing voltages to these loads does not always linearly reduce power consumption and can even have the opposite effect, especially if customer equipment voltages fall below industry guidelines. This effect is most pronounced for industrial customers with large induction motor loads, particularly for those that include adjustable speed drive mechanisms.

Apart from some regional pockets of activity, utilities have not embraced voltage reduction as a means of energy conservation. A more common practice is use of voltage reduction as an operational strategy related to the "operating reserve margin." This allows utilities to reduce voltage by 5% during critical peak periods (CPP) while remaining within preferred voltage standards. The most fundamental barrier to the consideration and adoption of voltage reduction for energy conservation objectives is technical skepticism over the link between voltage reduction and load reduction. Some utility engineers believe that certain loads draw more current at lower voltage levels and that, therefore, lower voltage does not necessarily result in reduced loads. However, while this inverse relationship may hold true for certain types of loads, data from many utilities clearly prove that most loads do consume less power at lower voltages within acceptable operating ranges.

Utilities cite several other reasons for skepticism about the potential effectiveness of voltage reduction. A "take-back effect" hypothesizes that voltage reductions will be only temporary because customers will adjust their usage based on perceived changes to the effect on end-use. For example, if lower voltages result in perceptibly dimmer lights, customers might simply switch to higher wattage bulbs, negating the expected energy savings.

Another cause for the utilities' concern is the increased risk of customer complaints stemming from lower voltage. Electronic devices such as computers and certain medical equipment are very sensitive even to minor voltage drops, which could cause life support and radiology equipment to fail, for example.

Long feeder lines present another problem for utilities serving rural areas, where occasional voltage drops below 114 V result in increased customer complaints. Even short feeder lines where customers have long secondary feeders result in large perceived voltage drops. In order to reduce these complaints, utilities need to invest in additional equipment such as capacitors or rework secondary systems to shorten the secondary conductors.

Utilities also regard their service territories and load characteristics as unique and tend not to share information about distribution voltage practices. Because most of the United States is not presently capacity-constrained, many utilities plan to reduce voltages only during system emergencies or for just a few peak days during the year.

Utilities also note that voltage regulation reduces power consumption, which poses an economic barrier for the companies. If a utility does not face peak capacity constraints or is unable to re-sell capacity on the wholesale market, it is hard-pressed to justify the economics of more stringent voltage regulation.

Provided that it can overcome its several obstacles, CVR promises to deliver energy more efficiently to end users.

DISTRIBUTION TRANSFORMER EFFICIENCY

Developments in transformer technology and construction can also improve the efficiency of power distribution. Three classes of single-phase and three-phase transformers are in common use on today's distribution systems: low-voltage, dry-type transformers for inputs less than or equal to 600 volts; medium-voltage, dry-type transformers for inputs of 601-34,500 volts; and liquid-immersed transformers for inputs of up to 2,500 kVA. Transformer efficiency is rated as a percentage when tested at a specified load. Following the standard U.S. test procedure TP-1 adopted by the National Electrical Manufacturers Association (NEMA) in 1996, low-voltage, dry-type transformers are rated at 35% of full load, and the other two transformer types are rated at 50% of full load.

Not all transformer classes exhibit the same levels of losses. Liquid-filled transformers are the most efficient, with losses of only about 0.25%, while medium-size dry-type transformer losses total about 5% and small dry-type unit losses equal about 2%. Aggregate transformer losses in U.S. power distribution can exceed 3%, or 140 billion kWh/year. The U.S. Environmental Protection Agency's Energy Star unit estimates that about 61 billion kWh/year of that power loss could be saved by using higher efficiency transformers (Global Energy Partners, 2005).

While transformers are generally very efficient, even a small amount of loss multiplied by the more than 25 million installed distribution transformers accounts for the largest single point of energy loss on power distribution systems.

Transformer Core and Winding Losses

Transformer losses consist of two types: core (or "no-load") losses and winding losses (also called "coil" or "load" losses). Core losses result from the magnetizing and de-magnetizing of the transformer core during normal operation; they do not vary with load, but occur whenever the core is energized (Fetters, 2002). Amorphous core transformers can reduce these core losses by as much as 80% compared with conventional materials (see Figure 2-3). Amorphous core transformers have not been widely adopted in the U.S. or Europe, due to a higher first cost than conventional transformers. However, demand is increasing rapidly in countries such as Japan, India, and China. Furthermore, the price differential is declining as silicon steel prices increase.

Winding losses occur when supplying power to a connected load. Winding loss is a function of the resistance of the winding material—copper or aluminum—and varies with the load. Conventional transformers use aluminum winding and are designed to operate at

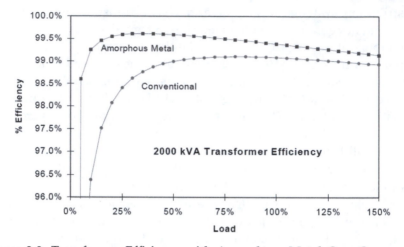

Figure 2-3. Transformer Efficiency with Amorphous Metal Core Compared with Conventional Steel Core. Source: Metglas.

temperatures up to 150°C/270°F above ambient. Newer high-efficiency transformers use copper winding, reducing the size of the core, the associated core losses, and the operating temperatures to 80°C or 115°C (145°F to 207°F) above ambient. Hence, overall transformer efficiency is lowest under light load, and highest at rated load, regardless of which core material is used.

New Transformer Efficiency Standards

On October 12, 2007, the U.S. Department of Energy (DOE) issued a final rule governing distribution transformer efficiency. The effective date of the new rule is November 13, 2007, and the standards it mandates will be applicable beginning on January 1, 2010. The standards apply to liquid-immersed and medium-voltage, dry-type distribution transformers. These new mandatory efficiency levels range higher than existing NEMA TP-1 standards (represented in the rulemaking process as TSL-1). In most cases, the standards range between TSL-4 (minimum lifecycle cost) and TSL-5 (maximum energy savings with no change in lifecycle cost) (see Figure 2-4) (Federal Register, 2007). The final rule notes that these levels are achievable using existing designs of distribution transformers.

DOE also concludes that the economic impacts on utilities of the new efficiency standards are positive. It points out in the final rule that

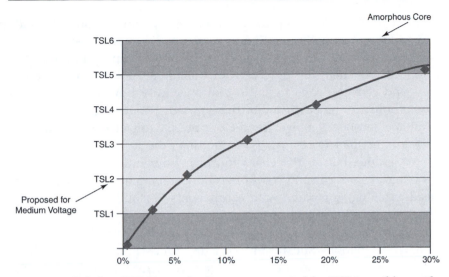

Figure 2-4. Existing NEMA standards are represented by TSL1 on this graph, while new DOE transformer efficiency standards range between TSL2—TSL3 and TSL4—TSL5 for transformers of varying voltages. The potential efficiencies of amorphous core transformers are at the high end of achievable efficiencies.

initial costs of liquid-immersed transformers will rise by 6 to 12%, but that operating costs (and electrical losses) will decrease by 15 to 23%. Similarly, the initial costs for medium-voltage, dry-type transformers will rise by 3 to 13%, but the corresponding losses and operating costs will decrease by 9 to 26%. On a national level, DOE projects that the standards will save about 2.74 quads of energy from 2010-2038, which is equivalent to the energy that 27 million homes consume in one year. This energy savings will decrease CO_2 emissions by about 238 million tons, which is equivalent to 80% of the emissions of all light vehicles in the U.S. for one year. Hence, DOE concludes that the standards are "technologically feasible and economically justified, and will result in significant energy savings (Federal Register, 2007)."

Advanced Transformer Technology
    The intelligent universal transformer (IUT), currently in the very early stages of development, is an advanced power electronic system concept that would entirely replace conventional distribution transformers. IUTs have a flat efficiency characteristic with a somewhat lower efficiency peak, but higher efficiency off-peak. They are expected to perform with less total loss (higher overall efficiency) over a daily load cycle.

The IUT would also provide numerous system operating benefits and added functionality relative to conventional transformers. An IUT is a solid-state transformer, similar to the power supply in a desktop computer. As such, the device would eliminate power quality problems and convert loads to sinusoidal and unity power factor, which would enhance the efficiency of the entire distribution grid. IUTs also eliminate secondary power faults, supply DC offset loads, and have very low no-load losses. At the same time, the IUT contains none of the hazardous liquid dielectrics found in conventional transformers, avoiding the hazards and costs of spills. Component costs for IUTs are steadily falling, compared with components in standard transformers, which are steadily rising in price.

IUTs can replace conventional distribution transformers, both in new installations and to replace aging units. Because the modules of IUT systems can be configured for several rating levels, IUTs can replace larger inventory requirements of many conventional transformers at different rating levels. Utilities can use IUTs as distribution system monitoring nodes to support system operations and advanced automation, and use optional functions (prioritized by sponsors) such as voltage regulation, configuring to supply three-phase power from a single-phase circuit, output ports for dc power and alternative ac frequencies, and interface with distributed generation.

While IUTs currently exist only in the laboratory, with active development and testing, models could be available within only a few years.

Advanced Distribution Automation (ADA)

The IUT is one component in a broader strategy called advanced distribution automation (ADA). ADA is distinct from traditional distribution automation (DA). Traditional DA enables automated control of basic distribution circuit switching functions. ADA is concerned with complete automation of all the controllable equipment and functions in the distribution system to improve strategic system operation. The various distribution system components are made interoperable in ADA, and communication and control capabilities are put in place to operate the system. The result is added functionality and improved performance, reliability, and cost, relative to today's system operations. In total, ADA will be a revolutionary change to distribution system infrastructure, as opposed to simple incremental improvements to DA.

While there are numerous areas where R&D advances are needed to realize ADA, five key areas include (Federal Register, 2007): system topologies (i.e., optimized radial and networked configurations); electronic/electrical technology development such as intelligent electronic devices (e.g., the IUT); sensor and monitoring systems; open and standardized communication architecture; and control systems.

Additional areas where significant development is required to achieve the objectives and the vision of ADA include:

- Development of an integrated distribution system with storage and distributed generation.

- Development of advanced tools to improve construction, troubleshooting and repair.

- Development and demonstration of a five-wire distribution system.

- Integration of smart metering concepts that would enable consumers and utilities to maximize the benefits of night-time recharging of plug-in electric vehicles.

CONCLUSION

The smart grid as a concept must extend from power production through delivery to end-use. Concepts of functionality employing sensors, communications and computational ability can be effectively used to reduce energy consumption, reduce emissions from power production, and improve reliability in the frontend of the electricity value chain.

References

Retrofitting Utility Power Plant Motors for Adjustable Speed: Field Test Program, M.J. Samotyj, Program Manager, EPRI Report CU-6914, December 1990.

Distribution Efficiency Initiative, Market Progress Evaluation Report, No. 1, Global Energy Partners, LLC, report number E05-139, May 18, 2005.

Fetters, John L., *Transformer Efficiency*, Electric Power Research Institute. April 23, 2002.

Energy Conservation Program for Commercial Equipment: Distribution Transformers Energy Conservation Standards; Final Rule. 10 CFR Part 431. Federal Register, Vol. 72, No. 197, 58189. October 12, 2007.

Smith, C.B., and K.E. Parmenter, "Electrical Energy Management in Buildings," in *CRC Handbook of Energy Conservation and Renewable Energy*, edited by F. Kreith and D.Y. Goswami, 2008.

Indoor Environmental Quality in Energy Efficient Homes: Marketing Tools for Utility Representatives, Global Energy Partners, LLC, Lafayette, CA: 2005. 1285-4-04.

Kempton, W. and L. Lutzenhiser, "Introduction," *Energy and Buildings*, vol. 18, no. 3 (1992), pp. 171-176.

Chapter 3

Electric End-use Energy Efficiency

DEFINING ELECTRIC END-USE ENERGY EFFICIENCY

A debate has raged for decades in the electric utility industry, centering on the issue of electric end-use energy efficiency as an alternative to traditional supply sources and to using fossil fuels at the point of end use. That debate now seems to be coming to closure. The utility industry is deeply rooted in the need for traditional, controllable sources of capacity and energy such as long-term contracts or power plants. Increasing costs, regulatory encouragement, and concerns about global warming have caused utility managers to consider demand-side activities. Demand-side planning involves those utility activities designed to influence customer use of electricity in ways that will produce desired changes in the utility's load shape, that is, changes in the pattern and magnitude of a utility's load. Energy efficiency as an alternative to traditional supply sources is no longer a debatable issue in the electric utility industry. As this debate has matured, the use of efficient electric end-use applications to displace fossil fuels has again surfaced as an essential part of an overall end-use efficiency strategy. In particular, it is the focus on the smart grid that enables this revival of interest.

ENERGY EFFICIENCY

Demand-side planning includes many load-shape-change activities including energy storage, interruptible loads, customer load control, dispersed generation, and energy efficiency. Energy efficiency involves a deliberate effort on behalf of the utility to promote change in the load shape (amount or pattern) by the customer through such hardware-related actions as improved building-thermal integrity coupled with increased appliance efficiencies through such non-hardware-related actions as altered consumer utilization patterns, as well as through the adoption of electric end-uses which displace fossil fuels while coincidentally reducing overall energy use and emissions.

Many individuals involved in the electric utility industry are uncomfortable when talking about energy efficiency. They are in the energy business and are dedicated to delivering kWh to consumers. Certainly few for-profit companies would readily embrace aggressive programs to decrease sales of their own products. Those supplying goods and services would have great difficulty embracing a role that involved spending time and money to convince customers that using fewer of their goods and services would be in the customer's best interest. However, these days, when utilities talk energy efficiency,* they often mean it, and many are convinced that they are doing so for good reasons. This reflects a desire to take a more holistic view of their customers' welfare, along with societal needs.

It has taken the world several decades to realize fully that energy is not an infinite resource. The illusion of infinite resources has supported and sustained a high standard of living. The electric sector has come to realize that a variety of energy sources and demand-side alternatives are necessary to maintain that standard of living, as well as to address global warming concerns and provide economic stability and national security. The energy industry has also come to realize that among all the measures that can be used as resources in meeting energy demand, energy efficiency is the simplest option and the most rewarding in the near term with regard to benefits to the energy supplier and the energy consumer.

Energy efficiency has the capability of saving all forms of non-renewable energy resources. While energy efficiency does seem somewhat inconsistent with the electricity "business" mission of selling energy, it may be beneficial for the supplier as well as for its customers and certainly for the environment.

IS ENERGY EFFICIENCY COST-EFFECTIVE?

But is it really cost-effective? Does it cost the consumer less in the long run?

---

*Based in part on material prepared by Clark W. Gellings and Patricia Hurtado, Kelly E. Parmenter and Cecilia Arzbacher of Global Energy Partners, LLC.

The question of whether energy efficiency is cost-effective depends on both the type of energy efficiency program under consideration and on the utility's characteristics. Energy efficiency programs and/or activities are utility-specific. Each must be evaluated on the basis of economics as applied to a particular utility. On some systems, energy efficiency may be the cheapest source of energy; however, each energy efficiency activity must be evaluated in light of capacity alternatives, environmental impacts and other energy efficiency options that are available. The overall economics will vary due to variations in weather, load characteristics, fuel costs, generation mix, etc. When a system benefit is not achieved to offset the cost of energy efficiency, the result may be higher costs.

In evaluating the benefits and costs of demand-side activities, one useful criteria which can be used is the impact on total costs. Since end-use activities can impact a variety of costs, evaluation usually is accomplished by a modeling technique that separately simulates the price of electricity to the average customer and to those participating in programs.

FINANCIAL IMPACTS OF ENERGY EFFICIENCY

Dealing with energy efficient technologies involves promoting the adoption of efficiency end-use technologies, the adoption of electric technologies which replace gas or oil end use, and/or invoking change in the customer's behavior. Costs and savings include a broad variety of items such as marketing, advertising and promotion, direct incentives, etc. In many cases, these form the greatest costs of an energy efficiency program. An additional economic benefit which is evolving entails using energy efficiency as a CO_2 credit. This can have huge impacts where efficient electric technologies displace inefficient gas or oil technologies.

HOW DESIRABLE IS ENERGY EFFICIENCY?

The desirability of embracing energy efficiency from an industry participant perspective depends on many factors. Among these are the fuel mix, capacity reserves, financial capability, and the potential CO_2 reductions from programs from the market participants in the region.

Wholesale market capacity\, prices, fuel mix and capacity purchases are the largest determinants of the marginal cost impacts of load-shape modification. It the marginal fuel is oil or gas for both on- and off-peak periods, then energy efficiency will offer great potential for cost reduction. If the fuel mix is such that off-peak marginal energy is produced by coal, nuclear, wind or hydropower, then other types of energy efficiency such as those involving load shifting, will be desirable.

A critical capacity situation provides a strong motive for interest in energy efficiency, especially if the region's ability to build new capacity is limited. In the case of investor-owned utilities, if stocks are selling below book value and the regulatory commission is not allowing an adequate rate of return, each new sale of stock hurts the existing shareholders. Energy efficiency may, therefore, benefit stakeholders by deferring the need for new capital. For distribution utilities without generation assets who are billed for capacity based on an inverted rate, the marginal benefits may exceed costs for energy efficiency programs.

On the other hand, where capacity is adequate or the ability to build new capacity is feasible, capacity (especially new capacity involving advanced coal or nuclear) may be cheaper than energy efficiency. However, neither of these options can be in place at the pace of energy efficiency. Since most regions are experiencing growth, it would be logical to approximate the cost of potential decreased units of energy by equating them to the marginal cost of supply additional increments of energy.

Inflationary pressures in many cases are forcing the marginal cost of electrical energy to exceed the average cost. For small changes in sales, it is logical to argue that the service provider could afford to use this difference (marginal less average) to promote or finance energy efficiency.

A RENEWED MANDATE

The 1970s and 1980s showed great strides in the rate of improvement of end-use electric energy efficiency. The oil embargos of the 1970s were a wake up call to much of the world that led to significant policy efforts to curb wasteful energy use and to develop and promote new energy efficient technologies, processes, and novel methods of energy supply, delivery, and utilization. The result was a notable increase in the efficiency of these production, delivery, and utilization technologies and systems and a companion decrease in the energy intensity of end-use devices and processes, which reduced the global rate of increase in energy consumption.

Since the 1990s, the global rate of efficiency improvement has slowed relative to the previous decades, with some regions experiencing lesser rates of improvement than others. Many factors have contributed to this deceleration; but, in essence, improvement in technology, the development of efficiency programs by utilities and government organizations, and the implementation of policies as a whole in the last one-and-a-half decades have not been as effective in furthering energy efficiency as in the 1970s and 1980s—the consequence of which is greater per-capita global demand for energy today than would have been required had efficiency gains kept up with the previous momentum. This discrepancy in efficiency improvement rates since the 1970s is an example of the unrealized potential of energy efficiency. This and other additional efficiency "resources" have yet to be fully tapped.

The increasing demand for electricity, the uncertainty in the availability of fossil fuels to meet tomorrow's energy needs, and concerns over the environmental impacts resulting from the combustion of those fuels—in particular, greenhouse gas emissions—combine to create a renewed urgency for energy efficiency gains. Policy makers and energy companies are well-positioned to develop strategies for meeting growing and changing energy needs of the public through energy efficiency.

Spurred on by both traditional and renewed drivers, energy efficiency is ready to make a resurgence with a new generation of technologies, policies and programs. History has demonstrated how the traditional drivers for energy efficiency have impacted worldwide energy consumption. The awareness of limited fossil fuel resources has created an impetus to improve the efficient production, delivery, and use of energy and develop cost-effective ways to implement renewable energy sources. This awareness has effected changes particularly in highly organized industrialized countries. Degradation to the environment from energy related causes has resulted in a significant international movement to protect the environment and its inhabitants. The increase in worldwide energy use that goes hand-in-hand with population growth and development has made it necessary to develop new technologies that can accomplish a given task with less energy. The threats to national security that result from dependence on non-domestic energy resources have resulted in policy efforts to maximize the use of domestic resources. Indeed, there has been a growing awareness of the value of a very important national and international resource—*energy efficiency*. This awareness is now heightened in the face of renewed drivers related to energy supply constraints, increasing fuel costs, and global efforts to reduce greenhouse gas emissions.

DRIVERS OF ENERGY EFFICIENCY

The primary reasons to increase worldwide energy efficiency can be grouped into four main factors:

1.   Worldwide energy consumption is growing due to population growth and increased energy use per capita in both developed and developing countries.

2.   Fossil fuel resources are finite, and the cost to extract and utilize these resources in an environmentally-benign manner is becoming increasingly expensive.

3.   A dependence on non-domestic energy supplies compromises national security for many nations.

4.   There is an increasing perception that the environment is suffering as a result of resource extraction, conversion, and utilization.

The relevance of each of these four factors is summarized as follows.

Increasing Worldwide Demand for Energy
The overall energy consumption of the world is increasing as a result of two main reasons. The primary reason is that the population is growing at an exponential rate. Since 1960, the world population has increased from 3.0 billion to 6.6 billion. At this rate, the population increases by about three people every second—for every five people born, two people die. The United Nations predicts that population growth will slow down and stabilize over the next century, due in part to increased family planning throughout the world. Even with a slowdown, it is estimated that the population will reach 9 billion by 2042. What does this mean for energy consumption? Each new person will need a share

of the world's energy resources. In addition, energy use per capita is increasing in both developed and developing countries. As countries become more industrialized, their success will depend on the level to which they implement efficient technologies and practices. For example, China is currently experiencing the most significant growth in energy demand due to both economic development and population growth.

Finite Resources

Much of the world depends on a finite supply of fossil fuels to meet their energy needs. The exact quantity of fossil fuel resources is under constant debate; although, the one point that cannot be disputed is that the expense to extract fossil fuels will become increasingly cost prohibitive as the supplies become more dilute and less available. The fuels that were once considered to be readily available and cheap are now becoming scarce and expensive. In addition, the perceived environmental impacts of fossil fuel combustion may be another limiting factor to their continued use. Some estimates predict that fossil fuels will only be viable as an energy source for one more century.

Other finite resources include nuclear fuels and unconventional hydrocarbons. The quantity of nuclear fuel resources for use in conventional light water reactors is thought to be less than fossil fuel reserves. There is a potential for significantly more useful energy to be extracted from nuclear reserves if the controversy surrounding nuclear power—in particular, the use of breeder reactors and the disposal of radioactive waste—is resolved. This application of nuclear fuels could extend resource availability by several hundred years or longer. Unconventional hydrocarbons (in particular, hydrogen) will likely be another resource of the future, but at the current time they are relatively uneconomical to produce and utilize.

Renewable resources have always been important and will increasingly be important in the future. However, further technical advancements are needed to improve the cost-effectiveness of replacing fossil fuels with renewable energy forms.

Energy efficiency is a relatively quick and effective way to minimize depletion of resources. It also buys time for the future development of alternative resources.

National Security

Fossil fuel resources are not evenly divided throughout the world.

In fact, resources for the two predominant energy sources, petroleum and natural gas, are concentrated in the Middle East and in Eastern European countries. Many of the countries in the Middle East are politically unstable. Several Eastern European countries are currently undergoing transition. Countries that rely on fuel imports from other, potentially less stable governments, run the risk of compromising national security. This is evidenced by continued tensions with the Middle East. Effective energy efficiency programs can reduce a country's reliance on non-domestic energy sources, which can in turn improve national security and stabilize energy prices.

The Environment

The environment appears to be experiencing damage as a direct result of industrialization and fossil fuel consumption. Gaseous and particulate emissions from power generation plants and many industrial processes impact the atmosphere at increasing rates. Effects to the environment include the destruction of forests in Central Europe by acid rain and potential climate change due to rising greenhouse gas levels. Moreover, exposure to air pollutants is hazardous to humans and other living species, and can cause a variety of health effects. In addition, the extraction and transportation of fossil fuels can result in environmental damage. For example, oil spills from tankers can be disastrous to the marine and coastal environments. The environment will greatly benefit from minimized use of environmentally hazardous energy forms, implementation of non-polluting renewable energy forms, increased air pollution control measures, waste management, and improved energy efficiency.

RENEWED INTEREST

The above considerations—population growth, increased energy use per capita in developing countries, limited resources, increased costs associated with extracting and converting less accessible energy forms, national security, and environmental degradation—are all important considerations that point to the need for improved energy efficiency. History has shown that energy efficiency works, and efforts that promote the development of energy efficient technologies and practices are effective alternatives to efforts that focus on the construction of new power plants and on the discovery and procurement of additional finite energy supplies. In fact, in the United States, energy efficiency has provided more "capacity" since the Arab oil embargo of 1973 than any efforts to increase the development of new resources.

The current energy state is in some ways very much like the state during the 1973 oil embargo. Many geographic regions have been experiencing rate shocks, there are still concerns over reducing dependence on foreign energy supplies from politically unstable regions, concerns that the environment is suffering, and of course, the worldwide demand for energy is still growing. Two of major differences seen today are 1) the increased cognizance that supplying peak loads in capacity constrained areas is very expensive, and 2) the greater concern that there may be a link between climate change and energy-related greenhouse gas emissions. Addressing the first issue has led to a proliferation of new demand response pilots and programs. (Section 5 discusses a new concept for marketing programs that integrates energy efficiency and demand response.) Addressing the second issue has resulted in a world-wide movement to reduce greenhouse gas emissions in hopes of arresting the pace of climate change.

The following subsection illustrates how energy efficiency is an important component in the portfolio of strategies to reduce greenhouse gas emissions.

Reducing Greenhouse Gas Emissions

The debate continues with some providing evidence they allege proves the anthropogenic (i.e., human-caused) forcing of climate change due to greenhouse gas emissions and others voicing out that there is insufficient reason to believe that recent climate change is not part of a natural cycle. It is not the point of this paper to get involved in the debate of whether or not anthropogenic greenhouse gas emissions are linked to climate change. Rather, the point is to show that reducing greenhouse gas emissions is, in fact, one of a number of motivations that support a global imperative to improve energy efficiency.

In terms of alternatives to reduce greenhouse gas emissions, experts believe that energy efficiency leads the list of a portfolio of strategies, which also include:

• Alteration of the power generation mix (e.g., greater use of renewables, nuclear power, and advanced coal);

- Fuel substitution (i.e., replacing existing use of greenhouse-gas-intensive fuels with cleaner energy forms in electricity generation, industrial processes, and transportation);

- Forestation;

- Carbon capture and storage (CCS); and

- Improvements in the energy efficiency of end-use devices and electricity generation, transmission and distribution.

Greenhouse gas mitigation strategies can be grouped into the seven main categories responsible for anthropogenic greenhouse gas emissions:

1. Energy supply and delivery
2. Transportation
3. Buildings
4. Industry
5. Agriculture
6. Forestry
7. Waste

Table 3-1 summarizes some of the primary strategies by sector. The technical maturity, implementation ease, and cost of the mitigation alternatives across the seven categories presented in Table 3-1 vary widely.

For the energy supply and delivery category, the most important mitigation alternatives consist of improvements in the efficiency of generation, transmission, and distribution; fuel switching; greater proportions of nuclear energy, renewables, and advanced coal technologies in the energy supply mix; and carbon capture and storage. Renewables are in the form of hydro-power, wind-power, bio-energy, geothermal energy, solar-thermal, and solar photovoltaic. Carbon capture and storage is general envisioned to be used in conjunction with gas- and coal-fired power plants.

The transport category includes strategies related to improving the fuel efficiency of on-road vehicles, marine vessels, rail transport, and aircraft. It also includes improving the operational efficiency (e.g., factors related to transport scheduling, passenger loading, etc.) of these modes of transport. Much focus has been placed on road transport because of its large share of mobile emissions. Measures particularly applicable to road transport include employing refrigerants with low global warm-

Table 3-1. Potential Strategies to Mitigate Greenhouse Gas Emissions (Metz, et al., 2007)

Energy Supply and Delivery	Transport	Buildings	Industry	Agriculture	Forestry	Waste
Generation efficiency	Fuel efficiency (road vehicles, aircraft, marine transport, rail transport)	Energy efficiency of building and systems	Energy efficiency of facilities and processes	Energy efficiency of buildings, transport, and systems	Maintain/increase forest area	Landfill CH_4 recovery and utilization
Transmission and distribution efficiency	Operational efficiency	Fuel switching	Fuel switching	Restoration of cultivated organic soils	Maintain/increase site-level carbon density	Improved landfill practices
Smart grid	Refrigerants with low global warming potential	Controlling non-CO_2 greenhouse gas emissions	Energy recovery	Cropland management	Maintain/increase landscape-scale carbon stocks	Wastewater management
Fuel switching	Biofuels		Renewables (e.g., biofuels, solar drying)	Grazing land management	Increase off-site carbon in wood products	Anaerobic digester biogas
Plant retirement	Hybrid vehicles		Recycled or scrap feedstocks and materials	Restoration of degraded lands	Increase bioenergy and fuel substitution	Controlled composting
New generation capacity	Hydrogen-powered fuel cell vehicles		Reduced losses	Rice management		Advanced incineration
- Nuclear	Electric vehicles		Controlling non-CO_2 greenhouse gas emissions	Livestock management		Expanded sanitation coverage
- Hydro	Modal shift to non-motorized transport		Carbon capture and storage	Biofuels		Waste minimization, recycling, and reuse
- Wind	Modal shift to public transport			Manure management		
- Bio-energy						
- Geothermal						
- Solar						
- Advanced coal						
Carbon capture and storage						

ing potential, using bio-fuels, or switching to hybrid vehicles, plug-in hybrid electric vehicles, hydrogen-powered fuel cell vehicles, or electric vehicles. Encouraging modal shifting from less to more efficient modes of transport is another alternative.

Mitigation strategies for residential and commercial buildings fall into three general areas: energy efficiency of buildings and systems, fuel switching, and controlling non-CO_2 greenhouse gas emissions. There are numerous potential energy efficiency measures ranging from those that address the building envelope to those that address systems such as lighting, water heating, space heating, space cooling, ventilation, refrigeration, and appliances.

The industrial category consists of measures to improve the efficiency of facilities and processes; fuel switching; energy recovery (e.g., combined heat and power); use of renewables; recycling and use of scrap materials; waste minimization; and controlling other (non-CO_2) greenhouse gas emissions. Carbon capture and storage is also potentially feasible for large industrial facilities.

Mitigation alternatives for the agriculture sector pertain to energy efficiency improvements and land, livestock, and manure management. The development and use of biofuels is a cross-cutting mitigation strategy that also relates to energy supply. In addition, the use of anaerobic digesters in manure management practices is also applicable to the general category of waste management.

Forestry strategies are aimed at either reducing emissions from sources or increasing removals by sinks. Specific alternatives include maintaining or increasing forest area, site-level carbon density, and/or landscape level carbon density; increasing off-site carbon stocks in wood products; bio-energy; and fuel substitution.

The category of waste management is comprised of technologies to reduce greenhouse gas emissions as well as to avoid emissions. Emission reduction approaches include recovery and utilization of methane (CH_4); landfill practice improvements; wastewater management; and generation and use of anaerobic digester gas. Emission avoidance methods consist of controlled composting; advanced incineration; expanded sanitation coverage; and waste minimization, recycling, and reuse.

Several recent studies have estimated the potential of these strategies in reducing greenhouse gas emissions, with some of the work taking into account the economic and market potentials and some of the work focusing just on the technical potential. For example, EPRI has analyzed several of these mitigation opportunities to assess the technical potential for future carbon dioxide (CO_2) emissions reductions, specifically, within the scope of the U.S. electricity sector. EPRI refers to this work as the "PRISM" analysis (see www.epri.com).

WHAT CAN BE ACCOMPLISHED?

Energy efficient end-use technologies and practices are some of the most cost-effective, near-term options for meeting future energy requirements. Many of them can be deployed faster and at lower cost than supply-side options such as new clean central power stations. Energy efficiency is also environmentally-responsible.

Several recent studies have attempted to estimate the potential for energy efficiency improvements. In addition, the costs associated with energy efficiency efforts have been compared with the costs of other strategies for meeting future energy needs in a clean, sustainable manner. Results indicate that energy efficiency gains have the potential to contribute significantly to meeting future energy requirements in a relatively cost-effective and easily-deployable manner.

A considerable amount of work has been undertaken in recent years to assess the energy efficiency potential. Some of the approaches rely on technologies that are already available; others count on expected technological advances (i.e., emerging technologies); and still others depend on accelerated technological developments or technology "leaps" to make alternatives viable. Many of the projections assume that mechanisms to remove technical, economic, and market barriers will be in place. The following subsections summarize some recent estimates of the future potential of energy efficiency, with a focus on electricity use.

IEA ESTIMATES

In their *World Energy Outlook 2006* (WEO), the International Energy Agency (IEA) discusses two potential scenarios for the world's energy future: the reference scenario and the alternative policy scenario (International Energy Outlook, 2006). The reference scenario accounts for all government energy and climate policies and measures enacted or adopted as of the middle of 2006 in the projection of future energy use patterns through 2030. This scenario does not take into account any future policies, measures, and/or new technologies. The alternative policy scenario on the other hand accounts for all energy and climate policies and measures currently being *considered*. Its projections, therefore, depict what is potentially possible in terms of reaching energy goals if we act right away to implement a set of policies and measures under consideration as of 2006. The types of new measures included in the alternative policy scenario include further improvements in energy efficiency, greater reliance on non-fossil fuels, and sustenance of the oil and gas supplies within net energy-importing countries. However, the Alternative policy scenarios not take into account technologies that have yet to be commercialized; rather it looks at the potential for greater improvement and increased and faster penetration of existing technologies compared to the reference scenario. Breakthrough technologies are not included.

Relative to the reference scenario, the policies and measures included in the Alternative policy scenarios paint a cleaner energy picture. The latter scenario is associated with significantly lower energy use and a greater share of less carbon-intensive energy sources.

In general, energy efficiency improvements in the Alternative policy scenarios correspond to a global rate of decrease in energy intensity (energy consumed per unit of gross domestic product) of 2.1% per year over 2004 to 2030 for the Alternative policy scenarios compared with 1.7% per year in the reference scenario. Much of this improvement over the reference scenario is in developing and transition economies where there is more potential for energy efficiency improvements.

In terms of electricity use, the global electricity demand (final consumption of electricity) in 2030 is estimated to be 12% lower in the Alternative policy scenarios (24,672 TWh) than in the reference scenario (28,093 TWh) mainly due to greater energy efficiency improvements (see Table 3 2). Two-thirds of the improvements come from energy efficiency in residential and commercial buildings (more efficient appliances, air conditioning, lighting, etc.) and one-third comes from improvements in the efficiency of industrial processes. The savings in residential and commercial buildings alone equates to avoiding 412 GW of new capacity, which is a little less than China's 2004 installed capacity. Still, compared with the electricity demand in 2004 (14,376 TWh), the electricity demand is estimated to increase by a factor of 1.7 by 2030 for the Alternative policy scenarios and by a factor of 2.0 for the reference scenario. Most of the increase in electricity demand is expected to be in developing countries.

Table 3-2. Comparison of Worldwide Electricity Demand in 2030 Projected by the WEO Reference Scenario and the Alternative Policy Scenario

2004 Electricity Demand	2030 Electricity Demand (Reference Scenario)	2030 Electricity Demand (Alternative Policy Scenario)	Difference between and Alternative Reference Policy Scenarios in 2030
14,376 TWh	28,093 TWh	24,672 TWh	3,421 TWh (12%)

Electricity demand is the final consumption of electricity.
Source: World Energy Outlook 2006, International Energy Agency, Paris, France: 2006.

Table 3-3 compares the electricity demand in 2030 of the two scenarios for selected regions. The largest percentage reductions relative to the reference scenario take place in Brazil, the European Union, China, and Latin America. In terms of the actual quantity of electricity savings, China leads with a savings of 814 TWh relative to the reference scenario.

UNITED NATIONS FOUNDATION ESTIMATES

In a 2007 report, the United Nations Foundation makes a more ambitious plan of action to improve energy efficiency. The plan entails "doubling" the global historic rate of energy efficiency improvement to 2.5% per year by G8 countries between 2012 and 2030. (UN Foundation, 2007 and [G8 countries include Canada, France, Germany, Italy, Japan, Russia, the United Kingdom, and the United States]). (For reference, the rate of energy efficiency improvement averaged 2% per year between 1973 and 1990 and then declined to an average of 0.9% per year between 1990 and 2004.) (International Energy Agency, 2006)* The plan also calls for G8 countries to reach out to +5 nations (Brazil, China, India, Mexico, and South Africa), specifically, as well as to other developing countries

---

*The historic energy efficiency improvement rates quoted apply to a group of 14 IEA countries: Austria, Canada, Denmark, Finland, France, Germany, Italy, Japan, the Netherlands, New Zealand, Norway, Sweden, the United Kingdom, and the United States.

Table 3-3. Comparison of Electricity Demand in 2030 Projected by the WEO Reference Scenario and the Alternative Policy Scenario, Selected Regions (International Energy Agency, 2006)

Region	2004 Electricity Demand [TWh]	2030 Electricity Demand (Reference Scenario) [TWh]	2030 Electricity Demand (Alternative Policy Scenario) [TWh]	Difference Between Reference and Alternative Policy Scenarios in 2030 [TWh]
United States	3,641	5,281	4,792	489 (9%)
Japan	965	1,175	1,047	128 (11%)
European Union	2,652	3,710	3,175	535 (14%)
Russia	58	954	872	82 (9%)
China	1,756	6,106	5,292	814 (13%)
India	442	1,733	1,570	163 (9%)
Latin America	698	1,640	1,419	221 (13%)
Brazil	349	651	547	104 (16%)
Middle East	477	1,221	1,082	139 (11%)
Africa	407	1,070	977	93 (9%)

Electricity demand is the final consumption of electricity.
Source: World Energy Outlook 2006, International Energy Agency, Paris, France: 2006.

to help them attain efficiency goals. The G8 countries represent 46% of global energy consumption and are economically best positioned to lead the way in addressing efficiency improvements that can then be followed by the developing countries. Together, the G8+5 account for roughly 70% of global primary energy use.

If accomplished, this rate of energy efficiency improvement would lower G8 total energy demand in 2030 by 22% relative to the IEA's reference scenario and would return energy use to close to 2004 values for these countries.

The rate of energy efficiency improvement proposed in the United Nations Foundation report is more aggressive than energy efficiency improvement rates referenced in other recent energy projection studies. For example, Figure 3-1 compares the United Nations Foundation rate of 2.5% with rates assumed in the IEA's Reference and Alternative Policy Scenarios.

The economics of the United Nations Foundation plan are favorable. The report estimates that to improve the rate of energy efficiency to a level of 2.5% per year an investment of $2.3 trillion would be required by the G8 countries ($3.2 trillion on a global basis) relative to the IEA's reference scenario. This level of energy efficiency would avoid $1.9 trillion in new energy supply by G8 countries ($3 trillion globally). Therefore, the net incremental investment required would be $400 bil-

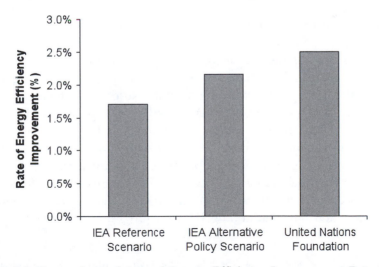

Figure 3-1. Comparison of Annual Energy Efficiency Improvement Rates (UN Foundation, 2007) Energy Efficiency Potential in the U.S.

lion by G8 countries ($200 billion globally). According to United Nations Foundation estimates, other quantifiable benefits such as reduced consumer energy bills help to pay back this relatively low investment in roughly three-to-five years (not including the value of climate change mitigation, energy security, etc.).

In addition to global-scale estimates, various entities have conducted assessments of the potential for energy efficiency at the country or regional level. For example, EPRI along with Global Energy Partners and The Keystone Center recently estimated the potential for end-use electric energy efficiency to yield energy savings and peak demand reductions (Gellings, et al., 2006 and Keystone, 2003). (Note: in this context, energy efficiency policies and programs are assumed to encompass traditional energy efficiency measures as well as demand response efforts.) The results of the study are summarized in the following paragraphs.

Table 3-4 displays the potential energy efficiency impacts pertaining to annual energy reduction in the year 2010. The analysis eliminated effects of energy efficiency programs currently underway or likely to occur according to historical trends. Therefore, it provides savings relative to a reference state of no energy efficiency program activity. It also reflects the *maximum* achievable potential at each cost level; the *realistic* achievable potential would likely be less. The total energy savings potential in 2010 is estimated to be close to 230 TWh, which is equivalent to 5.5% of the forecasted electricity use for 2010.* Significant impacts are associated with energy efficiency programs in the residential, commercial, and industrial sectors. A fifth of the potential is likely to be realized in the residential sector. Residential sector energy efficiency programs target high efficiency air conditioners that exceed current federal standards, improved building shell measures, and efficient lighting and appliances. A significant potential is likely to be achieved through improvements in end-use devices and enforcement of higher standards in new construction. The commercial sector is likely to offer the highest potential for energy saving opportunities (almost 50% of the total achievable potential is likely to be realized from the commercial sector). For this sector, energy efficiency programs target high efficiency heating, ventilation and air conditioning (HVAC) programs, improved lighting

---

*The forecasted total electricity use for 2010 is 4155 TWh according to *Annual Energy Outlook 2006, With Projections to 2030.* Energy Information Administration, Office of Integrated Analysis and Forecasting, U.S. Department of Energy, Washington, DC: February 2006.

Table 3-4. U.S. End-Use Electric Energy Savings Potential in 2010 (TWh)
Note EE is Energy Efficiency (Gellings, et al., 2006)

Residential Programs	Energy Savings (TWh)	Commercial Programs	Energy Savings (TWh)	Industrial Programs	Energy Savings (TWh)
New construction	10.6	New construction	35.8	Motors	24.1
Audits/weatherization	10.3	Lighting	33.5	Process EE	12.3
Lighting	6.3	Refrigeration	14.9	Lighting	17.9
Energy Star appliances	6.3	Cooling	13.4	Compressed air	5.0
HVAC tune-up/ maintenance	4.7	Equipment EE	12.6	Cooling	6.2
Air conditioning EE	4.5	Audits	5.4		
Refrigerators EE	2.0	Equipment tune-up/ maintenance	1.9		
Other equipment EE	1.8				
Fans EE	0.9				
Total Residential	**47.4**	**Total Commercial**	**117.5**	**Total Industrial**	**65.5**
				Total All Sectors	**230.4**

systems and high efficiency equipment such as refrigeration and motors. A very large part of this potential is likely to be realized from new construction activities. The remaining 30% of the potential is likely to be achieved from the industrial sector where energy efficiency programs target premium efficiency motors and efficiency improvements in manufacturing processes, as well as lighting improvements.

Table 3-5 displays the impacts pertaining to peak summer demand reductions for the year 2010, which is equivalent to 7.5% of the forecasted total electricity capacity for 2010.* Similar to energy savings, the residential sector is likely to contribute a fifth to the overall demand reduction impact. Residential sector peak summer demand impacts are represented by direct load control programs and energy efficiency programs. The commercial sector is forecasted to have the highest contribution to peak demand reductions (almost 42% of total demand reduction). Commercial sector peak summer demand impacts are represented by demand curtailment programs that serve to reduce loads during peak demand periods through the use of automated load control devices and energy efficiency programs, which yield the majority of the impacts mainly resulting from HVAC and lighting programs. The industrial sector demand reduction potential is close to that of commercial sector; peak summer demand impacts are driven primarily by demand curtailment programs and energy efficiency programs targeting motors and process uses.

Figure 3-2 is an energy efficiency supply curve constructed for this analysis. The supply curve shows the cost per kWh saved versus the amount of energy savings achievable at each level of cost. The curve is constructed by building up the savings from potential energy efficiency measures beginning with low-cost measures and continuing upward to high-cost measures. Analysis results show that savings of nearly 40 TWh can be achieved at a cost of less than $0.05/kWh. Similarly, savings of approximately 150 TWh are achievable for a cost of $0.10/kWh or less. In addition, savings of nearly 210 TWh can be achieved for less than about $0.20/kWh.

The low-cost portion of the supply curve offers many potential low-cost options for improving efficiency. Typical efficiency improve-

---

*The forecasted total electricity capacity for 2010 is 988.4 GW according to *Annual Energy Outlook 2006, With Projections to 2030*. Energy Information Administration, Office of Integrated Analysis and Forecasting, U.S. Department of Energy, Washington, DC: February 2006.

Table 3-5. U.S. Peak Summer Demand Reduction Potential in 2010 (GW)
Note EE is Energy Efficiency (Gellings, et al., 2006)

Residential Programs	Demand Reduction (GW)	Commercial Programs	Demand Reduction (GW)	Industrial Programs	Demand Reduction (GW))
Air conditioning direct load control	5.5	New construction	7.0	Demand response	13.9
Air conditioning EE	3.2	Demand response	5.9	Lighting EE	3.7
Water heating direct load control	2.4	Cooling EE	7.1	Motors EE	3.3
New construction	1.4	Lighting EE	6.9	Process EE	1.7
Audits/weatherization	1.3	Audits	1.1	Cooling EE	2.7
Lighting EE	0.7	Time-based tariffs	1.0	Time-based tariffs	1.6
Appliance removal	0.7	Refrigeration EE	1.3	Compressed air	0.6
Time-based tariffs	0.3	Equipment tune-up/ maintenance	0.3		
Energy Star appliances	0.2	Equipment EE	0.8		
Fans EE	0.2				
Other equipment EE	0.2				
Total Residential	16.1	Total Commercial	31.4	Total Industrial	27.5
				Total All Sectors	75.0

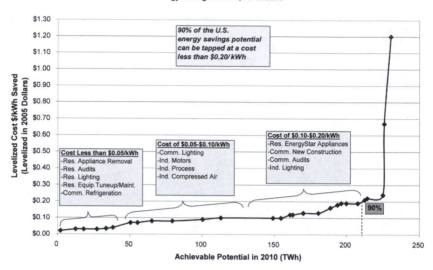

Figure 3-2. Supply Curve for U.S. End-Use Electric Energy Savings, 2010 (Gellings, et al., 2006)

ments include removal of outdated appliances, weatherization of the building shell, advanced refrigeration in commercial buildings, residential lighting improvements, and HVAC tune-ups and maintenance. All these technologies can be deployed for a cost of less than $0.05/kWh. For a cost of $0.05/kWh to $0.10/kWh, additional energy savings include commercial lighting improvements and efficiency improvements in industrial motors and drives as well as electro-technologies.

A similar analysis can be used to estimate the costs for reducing peak demand (Figure 3-3). Here, the appropriate metric is the cost of peak demand in units of $/kW, instead of $/kWh. Examples of peak demand savings opportunities that are available for less than $200/kW include audits and weatherization of residential building shells; commercial building tune-ups and maintenance; and time based tariffs for commercial and industrial customers. Other technologies that can be deployed at a cost of $200/kW to $400/kW include direct load control for residential air conditioning; advanced commercial (building-integrated) cooling and refrigeration systems; and advanced lighting systems for industrial, commercial, and residential applications.

The assessment shows that there is a substantial amount of cost-

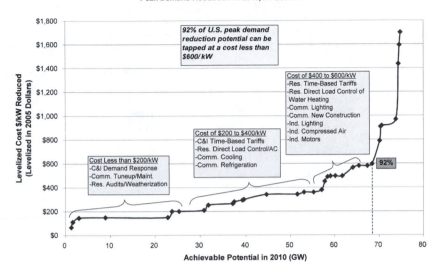

Figure 3-3. Supply Curve for U.S. Peak Demand Reduction, 2010 (Gellings, et al., 2006)

effective energy efficiency potential still achievable in the U.S. The opportunity for savings is highest in the residential and commercial sectors and is somewhat lower in the industrial sector.

CONCLUSION

Improving energy efficiency will require deliberate, concerted, and effective policies and programs at the international and local levels as well as extensive improvements in technology. All of the studies summarized here require that specific sets of policies and programs be implemented in order to maximize the potential for energy efficiency improvement. Each of these efforts is part of what should be considered as an overall approach to deploying a smart grid.

References
Testimony of C.W. Gellings, State of New Jersey—Board of Public Utilities—Appendix II, Group II Load Management Studies, January 1981.
C.W. Gellings, "Demand-side Planning," Edison Electric Institute Executive Symposium for Customer Service and Marketing Personnel, November 1982.

"Climate Change 2007: Mitigation," Contribution of Working Group III to the Fourth Assessment Report of the Intergovernmental Panel on Climate Change, B. Metz, O. R. Davidson, P. R. Bosch, R. Dave, L. A. Meyer (eds), Cambridge University Press, Cambridge, United Kingdom and New York, NY, USA: 2007.

World Energy Outlook 2006, International Energy Agency, Paris, France: 2006.

Realizing the Potential of Energy Efficiency, Targets, Policies, and Measures for G8 Countries, United Nations Foundation, Washington, DC: 2007.

Energy Use in the New Millennium: Trends in IEA Countries, International Energy Agency, Paris, France: 200

Gellings, C.W., Wikler, G., and Ghosh, D., "Assessment of U.S. Electric End-Use Energy Efficiency Potential," *The Electricity Journal*, Vol. 19, Issue 9, November 2006.

The Keystone Dialogue on Global Climate Change, Final Report, The Keystone Center, CO: May 2003.

Chapter 4

Using a Smart Grid to Evolve
The Perfect Power System*

Much has been written about optimizing power systems. Improvements are continually suggested and evaluated—some eventually implemented. But mostly the power system engineer's thinking is mired by two anchors: the existing system and the view that technical solutions regarding power systems are bounded by central generation on one end and the meter at the consumer's facility at the other.

The world's electricity infrastructure, consisting of various sources of generation, a set of transmission networks and a variety of distribution systems, has evolved through time. After electric utilities and power systems were created out of the need to supply arc lamps and electric streetcars in urban areas, the systems grew as true demand for electricity grew and as development spread. Throughout this expansion, the electrical system, and to a certain extent its design criteria, remain based on technologies that have been largely unchanged for 50 years.

Albeit the most complex machine ever built, it is largely comprised of simple parts—transformers, circuit breakers, cables and conductors. While some sensors have been added, the power system is bolted together, mechanically controlled and monitored by reactive computer simulation that is often 50 times slower than its operational movements.

As a result, electric service reliability to homes and businesses is far less than perfect. For example, over the last ten years every consumer, on average in most developed countries, has consistently been without power for 100 to 200 minutes or more each year. In addition, the quality of that service, even while on, has continued to deteriorate. There are now more perturbations in the quality of power than ever before. And

*The author wishes to acknowledge the contributions of Kurt E. Yeager, President Emeritus of the Electric Power Research Institute, and of the support of the Galvin Electricity Initiative to this section.

this at a time when consumers are increasingly using digital devices as society moves toward a knowledge-based economy.

Even less perfection is evident in the end-uses of the system—the energy consuming devices, appliances and systems which convert electricity into heat, light, motive power, refrigeration and the many other applications. Indeed to achieve perfection in electric energy service, engineers need to deploy the most efficient, environmentally friendly devices practicable and integrate them into building and processing systems that allow perfection to extend to the point of end-use.

So what if you could start with an absolutely clean sheet of paper—what if there was no power system and we could design one—a smart grid—using the best existing and evolving technology? This chapter takes precisely that perspective, to determine the path to the perfect system.

THE GALVIN VISION—A PERFECT POWER SYSTEM

The design of the perfect power system must start with the consumer's needs and provide absolute confidence, convenience and choice in the services provided so as to delight the consumer. Perfection, based on the consumer's perspective, must be the design principle. Traditional power system planners and practitioners will find this perfection concept hard to embrace. Delivering perfect electric energy service seems nearly impossible and appears inherently expensive—an impractical notion to those constrained by the conventional wisdom of the power sector.

In a project sponsored by the Galvin Electricity Initiative, Inc., the founders of Motorola Corporation have enabled researchers to develop a powerful vision, one that is unconventional, and capable of challenging the best minds in the industry. Bob Galvin, former CEO of Motorola and an industry icon, crystallized the framework that was to be developed by clarifying that:

> "My vision is not power system based on requirements being driven by future customer needs for end-use technologies, but a Perfect Power System unleashing self-organizing entrepreneurs whose innovations will greatly increase the value of electricity in the 21st century.

Further, he defined a clear and unambiguous measure of value—
"the system does not fail."

To define the path to a perfect electric energy service system, the
Galvin Initiative research team, in concert with the vision and inspiration
of Bob Galvin, literally took a "clean sheet of paper" to the challenge of
absolutely meeting the criteria of perfection.

As a result, the research team identified four potential system
Configurations associated with this path to the Perfect Power System:

- Perfect Device-level Power
- Building Integrated Power
- Distributed Power
- Fully Integrated Power: A Smart Grid

This path started with the notion that increasingly consumers
expect greater performance from end-use devices and appliances. Not
only does enhanced performance of end-use devices improve the value
of electricity; but also once perfection at this level is defined, it provides
elements of perfection that enable, in turn, a dispersed perfect system.
Dispersed perfect systems respond to consumer's demand for perfection,
but also accommodate increasing consumer demands for independence,
appearance, pride, environmentally friendly systems and cost control.

Dispersed systems or building integrated power systems can in
turn be interconnected to form distributed perfect systems. Once distributed perfect systems are achieved, they can be integrated with technologies that enable a fully integrated perfect power system to exist.

The research team used this broad array of technology development opportunities to establish general design criteria for the perfect
power system. This established the groundwork for subsequent tasks
and the eventual development of the possible system configurations and
associated nodes of innovation.

Defining the Perfect Electric Energy Service System

The research team used panels of energy experts to define a perfect
electric energy service system in terms of what it would accomplish:
*The Perfect Power System will ensure absolute and universal availability of
energy in the quantity and quality necessary to meet every consumer's needs.*
The focus of the system is on the service it provides to the energy users,
and any definition of a perfect system has to be considered from the

perspective of the energy users.

In order to provide service perfection to all energy users, the perfect power system must meet the following overarching goals:

- Be smart, self-sensing, secure, self-correcting, and self-healing

- Sustain failure of individual components without interrupting service

- Be able to focus on regional, specific area needs

- Be able to meet consumer needs at a reasonable cost with minimal resource utilization and minimal environmental impact (consumer needs include functionality, portability, and usefulness)

- Enhance the quality of life and improve economic productivity

Design Criteria

The design criteria that would be employed to meet these performance specifications must address the following key power system components or parameters:

1. End-Use Energy Service Devices
2. System Configuration and Asset Management
3. System Monitoring and Control
4. Resource Adequacy
5. Operations
6. Storage
7. Communications

The end-use devices are the starting point in the design of the perfect power system. They are the point of interface with the energy user and the mechanism by which the energy user receives the desired service, such as illumination, hot water, comfortable space conditioning (heating or cooling), and entertainment.

Path to the Perfect Power System

Now that the basic specifications and criteria for the perfect power system have been articulated, we can now focus on the various potentially perfect power system configurations envisioned by the Galvin Initiative. It is also important to recognize that nodes of innovation are required to enable and achieve perfection. In turn, key technologies are the basic building blocks needed to achieve the necessary innovative functionality advancements.

OVERVIEW OF THE
PERFECT POWER SYSTEM CONFIGURATIONS

The basic philosophy in developing the perfect power system is first to increase the independence, flexibility, and intelligence for optimization of energy use and energy management at the local level; and then to integrate local systems as necessary or justified for delivering perfect power supply and services.

This path started with the notion that increasingly consumers expect perfection in the end-use devices and appliances they have. Not only does portability enable a highly mobile digital society; but also once perfection in portability is defined, it provides elements of perfection that enable, in turn, a localized perfect system. Localized perfect systems can also accommodate increasing consumer demands for independence, convenience, appearance, environmentally friendly service and cost control.

Local systems can in turn be integrated into distributed perfect systems. Distributed perfect systems can, in turn be interconnected and integrated with technologies that ultimately enable a fully integrated perfect power system. Figure 4-1 summarizes each of these system configuration stages.

Each of these configurations can essentially be considered a possible structure for the perfect power system in its own right, but each stage logically evolves to the next stage based on the efficiencies, and quality or service value improvements to be attained. In effect, these potential system configuration stages build on each other starting from a portable power system connected to other portable power systems which then can evolve into a building integrated power system, a distributed power system and eventually to a fully integrated power system as reflected in Figure 4-2.

DEVICE–LEVEL POWER SYSTEM

The first level of development for the perfect power system is what we will call the "device-level" power system. This scenario takes advan-

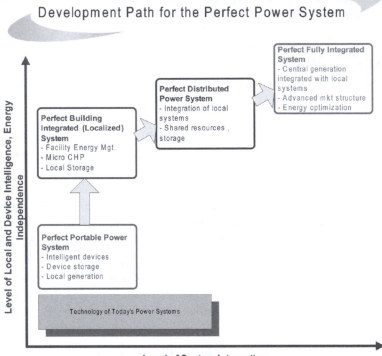

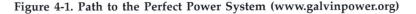

Figure 4-1. Path to the Perfect Power System (www.galvinpower.org)

tage of advancements in advanced technologies including nanotechnology, biosciences, sensors and advanced materials. This scenario has only modest needs for communication between different parts of the system as it essentially represents the capability of end-use technologies to operate in an isolated state and on their own with extremely convenient means of charging their storage systems from appropriate local energy sources.

Advantages of the Perfect Device-level
Power System and Relevant Nodes of Innovation

The perfect device-level power system has many advantages:

- The reliability and derivative quality of equipment and processes is determined locally and is, therefore, not dependent on a massive power delivery and generation infrastructure that has many possible failure scenarios.

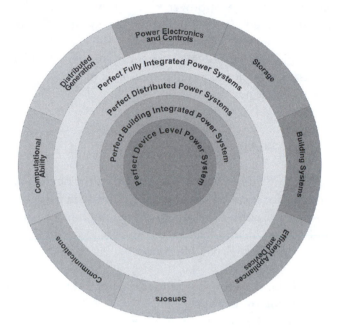

Figure 4-2. The Perfect Electric Energy System—System Configurations and Nodes of Innovation (www.galvinpower.org)

- It is the most flexible system configuration. The system is not dependent on any existing infrastructure (and, therefore, is ideal for developing systems and remote power requirements) but is also applicable to systems that are already developed (to take advantage of the existing infrastructure as an energy source for the device-level systems).

- Innovations in end-use technologies, storage, etc., can be utilized immediately without significant issues of control and integration with the power delivery system.

- There are tremendous opportunities for energy savings through local optimization of powering requirements and miniaturization of technologies.

- Renewables (solar, wind, other sources) can easily be integrated locally as an energy source for the device-level power systems.

Eventually device-level power systems may need to be connected to building integrated or distributed power systems for bulk energy supply.

An example of a device-level power system is provided in Figure 4-3. This illustrates various portable devices and inductive charging as elements of portable power.

BUILDING INTEGRATED POWER SYSTEMS

The "building integrated" power system is the next level of integration after the device-level power systems. In this scenario, energy sources and a power distribution infrastructure are integrated at the local level. This could be an industrial facility, a commercial building, a campus of buildings, or a residential neighborhood.

Advantages of the Building Integrated Power System & Relevant Nodes of Innovation

Integration at the local level provides a number of advantages over device-level power systems (at the cost of additional infrastructure and communication requirements needed for this integration):

• Energy availability can be optimized across a larger variety of energy sources, resulting in improved economics of power generation.

• Creates an infrastructure for more optimum management of overall energy requirements (heating, cooling, power) than is possible with the portable system.

• Still allows for local control and management of reliability at the local level.

One conceptual graphic for a building integrated power system is shown in Figure 4-4. The DC powered house combines many of the nodes of innovation required in the device-level power system and combines them into a single isolated power system such as a house or office complex.

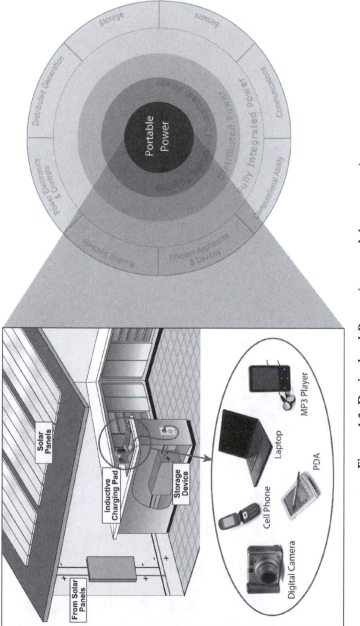

Figure 4-3. Device-level Power (www.galvinpower.org)

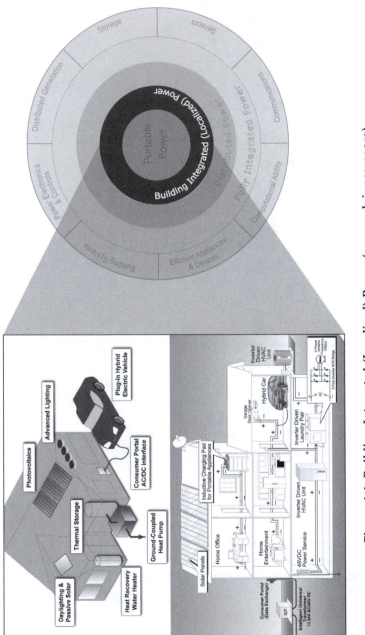

Figure 4-4. Building Integrated (Localized) Power (www.galvinpower.org)

DISTRIBUTED POWER SYSTEMS

The distributed power system involves interconnection of different localized systems to take advantage of power generation and storage that can support multiple local systems. This system still has limited needs for extensive power delivery grid infrastructures, but interconnection of local systems allows sharing of generation and storage capabilities over wider areas for more efficient energy management. The structure also can result in improved reliability by allowing for energy supply alternatives.

Advantages of the Distributed Power System & Relevant Nodes of Innovation

The main advantage of the distributed configuration is that it provides for additional flexibility in power generation and storage solutions. The main communications and control is still expected to be localized, and the energy management can be optimized at the local level, but also considering the availability of power from other sources besides the local systems. These more centralized systems can incorporate both generation and storage systems.

This additional flexibility is enhanced by a use of communications, control hierarchy and computational ability. This naturally results in an increase in infrastructure complexity. The concept of the distributed system is to optimize performance locally without complete dependence on the bulk power system infrastructure (this maximizes reliability), but being able to take advantage of the overall infrastructure to optimize energy efficiency and energy use.

In this configuration, energy performance can be optimized through a market structure where appropriate values are placed on all-important quantities (reliability, society costs of different sources of generation, importance of the energy to different devices and systems, storage systems availability, etc.). A real-time system can optimize both energy and reliability through the interconnection of the local systems, availability of central generation and storage, and reliability and power quality management technologies.

Additionally, reliability and power quality will be assured through technologies in individual devices and local systems, as well as centralized technologies. Like distributed generation, these power quality and reliability technologies will receive priority at the local level in this scenario with the added ability to export services or take advantage of some centralized services as appropriate to optimize overall performance. The DC distribution described in the localized system still provides a foundation for implementation of local technologies for improved quality, reliability and performance. Figure 4-5 provides a visual representation of a distributed configuration that integrates renewable and fuel cell technologies.

FULLY INTEGRATED POWER SYSTEM: THE SMART GRID

The final level of development involves a configuration that enables the complete integration of the power system across wide areas into a smart grid. The primary difference between this configuration and the distributed power configuration is the inclusion of a centralized generation sources and the possibly more modest use of distributed energy resources (DER), although the opportunity for DER to be included in the optimization is inherent in the design. The design implies full flexibility to transport power over long distances to optimize generation resources and the ability to deliver the power to load centers in the most efficient manner possible coupled with the strong backbone.

Figure 4-6 provides a simple diagram of a fully integrated perfect power system which integrates power electronics, building systems, sensors, communications and computational ability that provides for an optimal power system that is self healing and will not fail.

NODES OF INNOVATION

In addition to outlining the conceptual development of plausible perfect power system configurations it is equally important to identify "nodes of innovation" that are essential and contribute to the development of the perfect system configuration. These nodes can include the evolution of entirely new end-use devices, the proliferation of existing devices now available—but having limited market penetration today, plus any number of possible electric power systems technologies and configurations.

The Galvin Research Team selected the following nodes of in-

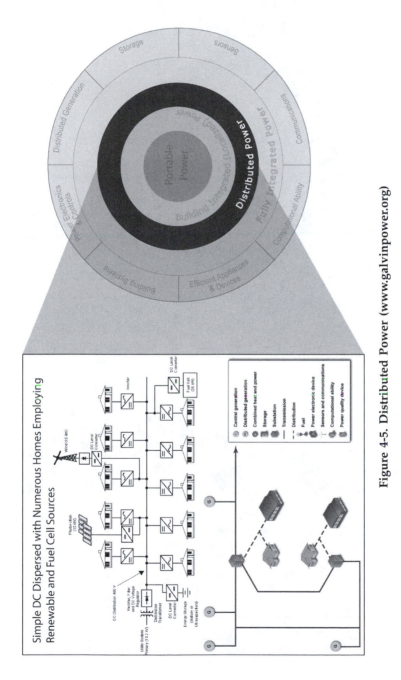

Figure 4-5. Distributed Power (www.galvinpower.org)

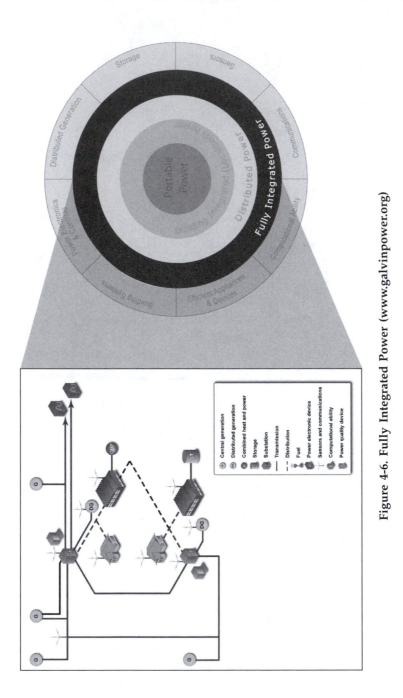

Figure 4-6. Fully Integrated Power (www.galvinpower.org)

novation that will be critical for developing the perfect electric energy systems described in the previous section. None of the perfect system configurations are capable of being deployed today without the evolution of one or more nodes of innovation being enhanced. These nodes can be most enhanced by certain elements of related technologies that may be part of those nodes.

The eight critical nodes of innovation identified include:

- Communications
- Computational Ability
- Distributed Generation
- Power Electronics and Controls
- Energy Storage
- Building Systems Efficient Appliances and Devices
- Sensors

The reader should note that there are a substantial number of key technologies that appear to have the greatest probability of advancing one or more of the candidate Perfect Systems. Future work is needed to study those technologies that may be essential to portable, dispersed or distributed configurations. It was this belief that once those systems had evolved, the perfect fully integrated configuration would literally self-organize.

For more information on the perfect power system visit the Galvin Electricity Initiative Website at www.galvinelectricity.org.

CONCLUSION

This chapter lays out possible configurations that may lead researchers and practitioners to stimulate thinking about evolving improved power systems resulting in a smart grid.

Power system engineers continue to debate the best approach to optimizing existing electric infrastructures in the developed world and planning and deploying new infrastructures in developing countries. Typically this debate is mired in a mental framework that evolves from the traditional electric industry. Starting from the beginning, engineers are taught to design, analyze and operate bulk power systems. Naturally, this approach is reinforced as graduate engineers enter industry. This chapter offers an entirely new approach. By taking a clean sheet of paper and approaching the design from a consumer perspective it is possible to unleash innovation and offer substantive improvement in quality and reliability. The result may well be the evolution of new systems—or it may be the augmentation of existing systems. Time will tell.

References

Phase I Report: The Path to the Perfect Power System, 2005, www. galvinpower.org.

Phase I Report: Technology Mapping, Scanning and Foresight, 2005, (www.galvinpower.org).

Chapter 5

DC Distribution & the Smart Grid

Thomas Edison's nineteenth-century electric distribution system relied on direct current (DC) power generation, delivery, and use. This pioneering system, however, turned out to be impractical and uneconomical, largely because in the 19th century, DC power generation was limited to a relatively low voltage potential and DC power could not be transmitted beyond a mile. Edison's power plants had to be local affairs, sited near the load, or the load had to be brought close to the generator.

Direct current (DC) is a continuous flow of electricity in one direction through a wire or conductor. Direct current is created by generators such as fuel cells or photovoltaic cells, and by static electricity, lightning, and batteries. It flows from a high to a low potential; for example, in a battery, from a positive to a negative pole. Any device that relies on batteries—a flashlight, a portable CD player, a laptop computer—operates on direct current. When represented graphically, DC voltage appears as a straight line, usually flat.

Alternating current (AC) is electricity that changes direction at regular intervals. It builds to a maximum voltage in one direction, decreases to zero, builds up to a maximum in the opposite direction, and then returns to zero once more. This complete sequence, or *cycle*, repeats, and the rate at which it repeats is called the *frequency* of the current. In the U.S., the AC power provided to a home outlet has a frequency of 60 cycles per second. This is expressed as 60 hertz (Hz), the hertz being a unit equal to one cycle per second.

AC VS. DC POWER: AN HISTORICAL PERSPECTIVE

Early power systems developed by Thomas Edison generated and delivered direct current (DC). However, DC power systems had many limitations, most notably that power typically could not be practically transmitted beyond a distance of about one mile.

Moreover, because changing the voltage of DC current was extremely inefficient, delivery of power with direct current in Edison's time meant that separate electric lines had to be installed to supply power to appliances and equipment of different voltages, an economically and physically impractical approach. Another limitation was that DC current incurred considerable power losses.

Edison's concept for electrification of the U.S.—which included royalties from his patents on direct current systems—was to deploy relatively small scale, individual DC plants to serve small areas—such as the Pearl Street Station, which powered a part of New York City's financial district.

But George Westinghouse's polyphase alternating current (AC) power system—invented by Nikola Tesla and used with transformers developed by William Stanley, Jr., who also worked for Westinghouse—proved to be far superior technically and economically. The voltage of AC could be stepped up or decreased to enable long distance power transmission and distribution to end-use equipment.

Edison fought vociferously against the use of alternating current-based systems, which he claimed would be dangerous because of the high voltage at which power would need to be transmitted over long distances.

"My personal desire would be to prohibit entirely the use of alternating currents. They are unnecessary as they are dangerous."
—*Thomas Edison, 1889, Scientific American*

He even went so far as to demonstrate the danger of AC by using it to electrocute a Coney Island elephant named Topsy who had killed three men. He also electrocuted numerous cats and dogs procured from neighborhood boys. But despite proving that alternating current could be an effective means of electrocuting these hapless creatures, its superiority to DC for transmission and distribution was compelling.

Alternating current can be produced by large generators, and the voltage of alternating current can be stepped up or down for transmission and delivery. The distance limitation of direct current and the difficulties of changing voltages proved critical factors in abandoning DC systems in favor of those based on AC. With DC systems, power had to be generated close to where it was used. This resulted in problematic reliability and economics. If the local plant failed, the entire system was down. And because initial power systems were devoted to lighting loads and systems only generated power at times of high usage, the cost of energy was high—often more than $1 per kWh when adjusted for inflation to present dollars (2005)—compared to an average cost for residential electricity today of 8.58¢ per kWh.

Engineers wanted to interconnect systems to improve reliability and overcome the economic limitations of DC electrical systems. If one area's power were out because of a problem at the generator, then the adjacent town would be available to pick up the load. In addition, by interconnecting isolated systems, a greater diversity of load was obtained, which would improve load factor and enable more economical operation of the generation plants.

Another major driver was the desire to make use of hydroelectric power sources located far from urban load centers, which made long distance transmission essential, and therefore made alternating current essential.

Transformers Transform the Power Delivery System

By using transformers, the voltage can be stepped up to high levels so that electricity can be distributed over long distances at low currents, and hence with low losses.

Transformers that could efficiently adjust voltage levels in different parts of the system and help minimize the inherent power losses associated with long-distance distribution were a critical enabling technology that led to today's AC-dominated power distribution system. Transformers do not work with DC power.

Effective transformers were first demonstrated in 1886 by William Stanley of the Westinghouse Company. According to the Institute of Electrical and Electronics Engineers (IEEE) History Center, Stanley first demonstrated the potential of transformers to enable AC transmission at Main Street in Great Barrington, Massachusetts.

He demonstrated their ability to both raise and lower voltage by stepping up the 500-volt output of a Siemens generator to 3000-volts, lighting a string of thirty series-connected 100-volt incandescent lamps, and then stepping the voltage back down to 500-volts. Wires were run from his "central" generating station along Main Street in Great Barrington, fastened to the elm trees that lined that thoroughfare. A total of six step-down transformers were located in the basements of some Main Street buildings to lower the distribution to 100-volts. A total of twenty business establishments were then lighted using incandescent lamps.

Stanley's demonstration of raising the generator voltage to 3000-volts and then back down again was exactly the same concept as employed in present day power systems where a "generator step-up" transformer is used to raise the system voltage to a very high level for long distance transmission, and then "large substation" transformers are used to lower the voltage to some intermediate level for local distribution.

Similar alternating current systems that use transformers eventually replaced Thomas Edison's direct current systems. Stanley's installation in Great Barrington was the first such system to include all of the basic features of large electric power systems as they still exist more than one hundred years later.

Centralization Dictates AC Instead of DC

Other factors led to the preference for AC power transmission instead of DC power delivery—most notably a desire for large-area grids relying on centralized power plant, such as hydroelectric dams. Having a transmission and distribution system that could provide hydro-electricity to cities or to remotely located industries such as gold or silver mines in the Rocky Mountains was also an economic imperative.

Such development relied not only on transformers, but on development of polyphase alternating current generators per the Institute of Electrical and Electronics Engineers (IEEE).

Niagara Falls represented a showplace of a very different sort. Here electrical engineers were confronted with one of the great technical challenges of the age—how to harness the enormous power latent in Niagara's thundering waters and make it available for useful work. Years of study and heated debate preceded the start-up of the first Niagara Falls Power Station in the summer of 1895, as engineers and financiers argued about whether electricity could be relied on to transmit large amounts of power the 20 miles to Buffalo and, if so, whether it should be direct or alternating current. The success of the giant polyphase alternating current generators made clear the directions that electric power technology would take in the new century.

In the 25 years following the construction of the Niagara Falls Power Station, various technological innovations and other factors led away from the early small-scale DC systems, and toward systems based upon increasingly larger-scale central-station plants interconnected via transmission lines that carried alternating current. Now cities and towns could be interconnected, and power could be shared between areas. During this period, transmission voltages as high as 150 kV were being introduced, and so relatively large amounts of power could be transmitted efficiently over long distances.

In addition to technical and market forces, the government also played a role in development of centralized power systems and thus reliance on AC transmission. Public policy and legislation encouraged the movement to larger centralized systems, and by the 1970s, more than 95% of all electric power being sold in the U.S. was through large centralized power systems.

As a result, alternating current (AC) distribution was far superior for the needs of a robust electrical infrastructure. Unlike DC power, the voltage of AC could be stepped up with relatively simple transformer devices for distance transmission and subsequently stepped down for delivery to appliances and equipment in the home or factory. And Nikola Tesla's invention of a relatively simple AC induction motor meant end users needed AC, which could be generated at large central plants for high-voltage bulk delivery over long distances. (See the section *AC versus DC: An Historical Perspective* for more on the attributes of AC and DC power, and why AC originally prevailed.)

Despite a vigorous campaign against the adoption of alternating current, Edison could not overcome the shortcomings of his DC system. AC won out, and today utilities generate, transmit, and deliver electricity in the form of alternating current.

Although high-voltage direct current (HVDC) is now a viable means of long-distance power transmission and is used in nearly a 100 applications worldwide, no one is advocating a wholesale change of the infrastructure from AC to DC, as this would be wildly impractical.

But a new debate is arising over AC versus DC: should DC power delivery systems displace or augment the AC distribution system in buildings or other small, distributed applications? Edison's original vision for a system that has DC generation, power delivery, and end-use loads may come to fruition—at least for some types of installations. Facilities such as data centers, campus-like groups of buildings, or building sub-systems may find a compelling value proposition in using DC power.

Several converging factors have spurred the recent interest in DC power delivery. One of the most important is that an increasing number of microprocessor-based electronic devices use DC power internally, converted inside the device from standard AC supply. Another factor is that new distributed resources such as solar photovoltaic (PV) arrays and fuel cells produce DC power; and batteries and other technologies store it. So why not a DC power distribution system as well? Why not eliminate the equipment that converts DC power to AC for distribution, then back again to DC at the appliance?

Advocates point to greater efficiency and reliability from a DC power delivery system. Eliminating the need for multiple conversions could potentially prevent energy losses of up to 35%. Less waste heat and a less complicated conversion system could also potentially translate into lower maintenance requirements, longer-lived system components, and lower operating costs.

In a larger context, deployment of DC power delivery systems as part of AC/DC "hybrid" buildings—or as a DC power micro-grid "island" that can operate independently of the bulk power grid—could enhance the reliability and security of the electric power system.

BENEFITS AND DRIVERS OF DC POWER DELIVERY SYSTEMS

Due, in part, to the interest in the smart grid, the specter of several potential benefits are driving newfound interest in DC power delivery systems:

Increasingly, equipment operates on DC, requiring conversion from AC sources. All microprocessors require direct current and many devices operate internally on DC power since it can be precisely regulated for sensitive components. Building electrical systems are fed with AC that is converted to DC at every fluorescent ballast, computer system power supply, and other electronic device. As one specialty electronics manufacturer put it, "DC is the blood of electronics."

> AC-DC conversions within these devices waste power. The power supplies that convert high-voltage AC power into the low-voltage DC power needed by the electronic equipment used in commercial buildings and data centers typically operate at roughly 65% to 75% efficiency, meaning that 25 to 35% of all the energy consumed is wasted. About half the losses are from AC to DC conversions, the rest from stepping down DC voltage in DC to DC conversions.

Simply getting rid of the losses from AC to DC conversion could reduce energy losses by about 10 to 20%. Likewise, an increasing number of portable gadgets such as cell phones and personal digital assistants (PDAs) require an AC-DC adapter, which also results in power losses during conversion. Considered in aggregate, the millions of AC to DC conversions necessitated for the operation of electronics extract a huge energy loss penalty.

Distributed generation systems produce DC power. Many distributed generation sources such as photovoltaic cells and fuel cells—and advanced energy storage systems (batteries, flywheels, and ultra capacitors) produce energy in the form of DC power. Other devices can also be suited to DC output, such as microturbines and wind turbines. Even hybrid vehicles such as the Toyota Prius could serve as DC generators in emergencies with the right equipment to connect them to the electrical system.

The energy losses entailed in converting DC to AC power for distribution could be eliminated with DC power delivery, enhancing efficiency and reliability and system cost-effectiveness. For instance, EPRI Solutions estimated that the total lifecycle cost of PV energy for certain DC applications could be reduced by more than 25% compared to using a conventional DC to AC approach—assuming that the specific end-use applications are carefully selected.[2] The costs of new distributed generation such as PV arrays are still high, so optimization of designs with DC power delivery may help spur adoption and efficient operation.

Storage devices such as batteries, flywheels and capacitors store and deliver DC power. This again helps avoid unnecessary conversions between AC and DC.

DC power could help power hybrid automobiles, transit buses, and commercial fleets. Plug-in hybrid vehicles can go greater distances on electricity than today's hybrids since they have larger batteries. These batteries store DC power, so charging them with electricity from solar photovoltaic arrays and other distributed sources could reduce reliance on gasoline, enhancing security and emergency preparedness.

DC power delivery could potentially enhance energy efficiency in data centers, a pressing need. One of the most promising potential applications of DC power delivery is in data centers, which have densely packed racks of servers that use DC power. In such centers, AC is converted to DC at the uninterruptible power supply, to facilitate storage, then is converted again to push out to the servers, and is converted one more time to DC at each individual server. These conversions waste power and generate considerable heat, which must be removed by air conditioning systems, resulting in high electricity costs.

In a 10–15 megawatt (MW) data center, as much as 2–3 MW may be lost because of power conversions. As these centers install ever more dense configurations of server racks, DC power delivery systems may be a means to reduce skyrocketing power needs.

Improved inverters and power electronics allow DC power to be converted easily and efficiently to AC power and to different voltage levels. Component improvements enable greater efficiency than in the past, and improve the economics of hybrid AC/DC systems. Although improved electronics also enhance AC-only systems, such enabling technology makes the DC power delivery option feasible as well.

The evolution of central power architecture in computers and other equipment simplifies DC power delivery systems. At present, delivering DC to a computer requires input at multiple voltages to satisfy the power needs of various internal components (RAM, processor, etc.) Development of a central power architecture, now underway, will enable input of one standardized DC voltage at the port, streamlining delivery system design.

DC power delivery may enhance micro-grid system integration, operation, and performance. A number of attributes make DC power delivery appealing for use in micro-grids. With DC distribution, solid-state switching can quickly interrupt faults, making for better reliability and power quality. If tied into the AC transmission system, a DC power micro-grid makes it easy to avoid back-feeding surplus generation and fault contributions into the bulk utility system (by the use of a rectifier that only allows one-way power flow). In addition, in a low-voltage DC system, such as would be suitable for a home or group of homes, a line of a given voltage rating can transmit much more DC power than AC power.

Of course, while DC circuits are widely used in energy-consuming devices and appliances, DC power delivery systems are not commonplace, and therefore face the obstacles any new system design or technology must overcome. For any of the benefits outlined above to be realized, testing, development, and demonstration are needed to determine the true potential and market readiness of DC power delivery.

POWERING EQUIPMENT AND APPLIANCES WITH DC

Many energy-consuming devices and appliances operate internally on DC power, in part because DC can be precisely regulated for sensitive components. An increasing number of devices consume DC, including computers, lighting ballasts, televisions, and set top boxes. Moreover, if motors for heating, ventilating, and air conditioning (HVAC) are operated by variable frequency drives (VFD), which have internal DC buses, then HVAC systems that use VFDs could operate on DC power. Numerous portable devices like cell phones and PDAs also require an AC-DC adapter. As discussed above, by some estimates the AC-DC conversions for these devices waste up to 20% of the total power consumed.

Equipment Compatibility
EPRI Solutions examined the compatibility of some common devices with DC power delivery in 2002:

• Switched mode power supplies, including those for computers (lab test)
• Fluorescent lighting with electronic ballasts
• Compact fluorescent lamps (lab test)
• Electric baseboard and water heating units
• Uninterruptible power supplies (UPS)
• Adjustable speed motor drives

These devices represent a large percentage of the electric load, and EPRI Solutions' preliminary assessments show that each could be potentially powered by a DC supply. Although additional testing is needed to determine the effect of DC power on the long-term operation of such equipment, results do indicate the feasibility of delivering DC power to these devices.

Switched-mode Power Supply (SMPS)—Switched-mode power supply (SMPS) technology is used to convert AC 120 V/60 Hz into the

DC power used internally by many electronic devices. At the most basic level, an SMPS is a high frequency DC-DC converter.

Many opportunities exist to use DC power with SMPS-equipped equipment since SMPS technology is found in many electronic devices including desktop computers, laptop computers with power adapters, fluorescent lighting ballasts, television sets, fax machines, photocopiers, and video equipment. Although AC input voltage is specified for most of the electronic devices that have SMPS, in some cases, this equipment can operate with DC power without any modification whatsoever. Also, in many instances, the location on the SMPS where AC is normally fed could be replaced with DC.

Power Supplies for Desktop and Laptop Units—According to research on power supply efficiency sponsored by the U.S. Environmental Protection Agency and the California Energy Commission, as of 2004, there were nearly 2.5 billion electrical products containing power supplies in use in the U.S., with about 400 to 500 million new power supplies sold each year.

The total amount of electricity that flowed through these power supplies in 2004 was more than 207 billion kWh, or about 6% of the national electric bill. Researchers determined that more efficient designs could save an expected 15 to 20% of that energy. That amount represents 32 billion kWh/year, or a savings of $2.5 billion. If powered by DC, conversion losses could be reduced and significant savings achieved.

EPRI Solutions conducted tests to assess the ability of two standard SMPS-type computer power supplies to operate on DC; a 250 watt ATX type typically used for desktop computers, and a portable plug-in power module for laptop computers. In both cases, tests revealed that the power supplies would operate properly when supplied with DC power of the right magnitude, although no tests were done to determine power supply operation and performance when connected to the computer loads.

For the desktop computer power supply, sufficient output was provided when supplied with 150 V of DC or greater. For the laptop SMPS unit, 30 V DC was required to "turn on" the output, which begins at 19.79 V and continues at that output unless DC supply drops to 20 volts DC or below.

Fluorescent Lighting with Electronic Ballasts—The key to DC operation of fluorescent lights lies in the use of electronic ballasts. The ballast is used to initiate discharge and regulate current flow in the lamp. Modern electronic ballasts function in much the same manner as a switched-mode power supply thus making it potentially possible to operate them from a DC supply.

Virtually all new office lighting systems use electronic ballasts, which are more efficient and capable of powering various lights at lower costs. Only older installations are likely to have the less-efficient magnetic ballasts in place.

For electronically ballasted applications, several manufacturers make ballasts rated for DC. Lighting systems could be retrofitted with DC-rated ballast units for DC operation. All light switches and upstream protection in line with DC current flow would also need to be rated for DC.

Compact Fluorescent Lamps—Compact fluorescent lamps (CFL) are energy-efficient alternatives to the common incandescent bulb. A new 20-watt compact fluorescent lamp gives the same light output as a standard 75-watt incandescent light bulb, and also offers an average operating life 6 to 10 times longer.

A compact fluorescent lamp has two parts: a small, folded gas-filled tube and a built-in electronic ballast. As with the fluorescent tubes used in commercial lighting, the electronic ballast enables DC operation of CFLs. EPRI Solutions' testing of a 20-watt CFL unit with DC power supply revealed that while the CFL could operate on DC power, it required a much higher DC input voltage. With AC supply, the CFL provided constant light at 63 V, but with DC supply, 164.4 V DC was required.

After speaking with CFL manufacturers, EPRI Solutions researchers determined that the CFL used a voltage doubling circuit on the input to the electronic ballast. However, the voltage doubling circuit does not operate on DC voltage. Hence, the DC voltage must be twice the magnitude of the AC voltage to compensate for the non-functioning doubling circuit. This resulting over-voltage on the capacitors could result in shortened lamp life, depending on the ratings of certain input elements in the circuit.

The reduction in lamp life is unknown. Additional research is needed to determine whether the energy savings over the life of the lamp would compensate for the increased cost due to premature lamp failure.

Electric Baseboard and Water Heating—DC voltage can be used to run almost any device utilizing an electric heating element, including resistive baseboard and electric water heaters. In these applications, electrical current flowing in a heating element produces heat due to resistance.

The chief concern of using DC in such applications is not in the heating element itself, but in the contactors, switches, and circuit breakers used for such circuits. Since DC is more difficult to interrupt, the interrupting devices must be capable of clearing any faults that develop. There are no DC equivalents to ground fault circuit interrupters (GFCIs), which are commonplace electrical devices used in AC systems to prevent electric shock.

Uninterruptible Power Systems—Uninterruptible power systems (UPS) are excellent candidates for DC power support. A UPS is composed of an inverter, a high-speed static switch, various controls, and battery energy storage. The functional objective of the UPS is to provide high reliability and power quality for connected loads that may be susceptible to voltage sags or short duration power interruptions. Most UPS systems have anywhere from a few minutes up to about 30 minutes of battery storage. For larger UPS units (>30 kVA), it is typical to have a backup generator that starts and picks up load a few minutes after the utility power is interrupted, which, today, is lower cost than having several hours of battery energy storage onsite.

Since a UPS has an inverter and an internal DC bus, it already has many of the elements needed to operate with DC energy.

Variable speed motors—Motors are very important electrical devices, and represent a significant portion of power use in the U.S. In industry, for instance, approximately two-thirds of the electricity use is attributable to motors.

Most AC motor loads still use the same basic technology as the Tesla induction motor. These omnipresent motors convert AC power for applications such as air handling, air compression, refrigeration, air-conditioning, ventilation fans, pumping, machine tools, and more.

A workhorse of modern society, these motors can only operate with AC power. In fact, if subjected to DC power, an AC motor could burn up quickly. In addition, without alternating current, the magnetic vectors produced in the induction motor powered with DC would not be conducive to rotation and the motor would stall—so an induction motor simply will not operate directly on DC power.

But DC can be used if a variable frequency drive is part of the system. A variable frequency drive allows for adjusting the motor speed, rather than operating it either on or off. By varying the frequency of power over a wide range, motor speed can be adjusted to best match the mechanical process, such as circulating air with a fan. This ability to adjust speed can translate into significant energy savings, as a CEO for a major manufacturer explains:

Since a variable frequency drive converts 60 Hz power to DC and then converts the DC to variable frequency AC that is fed to the motor, a DC supply can be readily accommodated, further increasing energy efficiency.

Greater adoption of energy-efficient variable speed motors, now underway for heating, ventilating and air conditioning systems and other applications, represents a greater opportunity for deploying DC power. In addition, several manufacturers now offer DC variable frequency drives for solar-powered water and irrigation pumps.

DATA CENTERS AND INFORMATION
TECHNOLOGY (IT) LOADS

One of the nearest-term applications for DC power delivery systems is data centers, or "server farms." These facilities are strong candidates for DC power delivery due to: (1) the availability of products that could enable near-term implementation; and (2) an economic imperative to increase energy efficiency and power reliability.

A data center may consist of thousands of racks housing multiple servers and computing devices. The density of these servers keeps increasing, wasting power and generating heat with multiple AC to DC conversions.

The need to provide more and more power to new blade server technology and other high-density computing devices has made reducing electricity costs a pressing goal within the data center industry. Multiple approaches are under consideration to increase energy efficiency, including a multi-core approach, with cores running at reduced speed, and software that enables managers to run multiple operating system images on a single machine. However, one of the more intriguing options is DC power delivery. In fact, a data-center industry group formed in late 2005 with support from the California Energy Commission through the Lawrence Berkeley National Laboratory is exploring the challenge of determining how DC power delivery systems can reduce energy needs and enhance the performance of data centers.

Headed by the Lawrence Berkeley National Laboratory and implemented by EPRI Solutions and Ecos Consulting, the group has obtained funding from the Public Interest Energy Research (PIER), the California Energy Commission (CEC) and the California Institute for Energy Efficiency (CIEE) for a DC demonstration project at a Sun Microsystems facility in Newark, California. The objectives of the demonstration are to show:

1.   How DC-powered servers and server racks can be built and operated from existing components.

2.   The level of functionality and computing performance when compared to similarly configured and operated servers and racks containing AC power supplies.

3.   Efficiency gains from the elimination of multiple conversion steps in the delivery of DC power to server hardware.

Numerous Silicon Valley giants including Intel, Cisco, and others are participating and contributing to the project, including Alindeska Electrical Contractors, Baldwin Technologies, CCG Facility Integration, Cingular Wireless, Dranetz-BMI, Dupont Fabros, EDG2, Inc., EYP Mission Critical, Hewlett-Packard, Liebert Corporation, Morrison Hershfield Corporation, NTT Facilities, Nextek Power, Pentadyne, RTKL, SBC Global, SatCon Power Systems, Square D/Schneider Electric, Sun Microsystems, TDI Power, Universal Electric Corp., and Verizon Wireless.

The existing AC-based powering architecture in a data center, which requires multiple AC-DC-AC conversions, can have an overall system efficiency lower than 50%. How much energy and money could be saved by eliminating these multiple conversions? Field performance data are yet to be documented. However, preliminary estimates of energy savings indicate that about 20% savings could be realized by changing from AC-based powering architecture to DC-based powering architecture for a rack of servers. Table 5-1 shows one estimate from EPRI, which indicates that a typical data center of 1,000 racks could save $3.5 million annually by using a DC power delivery system.

To calculate energy savings estimates for different design configurations or using different assumptions, visit an Excel-based calculator, available at the Lawrence Berkeley National Laboratory website (http://

Table 5-1a. Energy savings estimate for one rack of servers with high-efficiency power conversion

	Total input power (Watts)	Reduction due to air conditioning (Watts)	Total savings (Watts)	Yearly energy savings (MWh)	Yearly energy savings ($)	Net present value (NPV) of savings ($)
AC power*	8,590					
DC power	6,137					
Savings	2,453	837	3,290	28.82	$3,458	$11,984

*The efficiencies for the AC system are based on typical, rather than best-in-class systems. If a best-in-class AC system is compared to a DC best-in-class system, the savings from use of DC power would be reduced. For instance, yearly energy savings might be about $873 rather than $3428. However, gains in reliability from DC power (not shown in this table) would not be achieved.

Only energy-related savings are considered; other savings such as size and heat sink cost not considered. Calculations are based on typical power budget for a dual 2.4 GHz Xeon processor based 1U server rack

1U = TK

Energy cost = 12¢/kWh; project life = 4 years; discount rate = 6%; Overall cooling system efficiency = 1,200 Watts/ton; number of 1U servers per rack = 40

Table 5-1b. Assumptions

Power conversion efficiency	AC power architecture	DC power architecture
UPS	85%	N/A
AC/DC PS	72%	N/A
DC/DC VRM (12V – 1.75 V)	84%	84%
DC/DC (48V-12V)	N/A	95%
Nextek power module	N/A	92%
1U dual processor server power budget	**Typical (W)**	**Maximum (W)**
Dual processor power (@1.75V DC)	60	130
Mother board, PCI Card, DDR memory and other peripheral DC power consumption (@12V, 5V, and 3.3V DC)	60	220

hightech.lbl.gov/DC-server-arch-tool.html).

Intel has estimated that power consumption can be reduced by about 10%, and others have projected even higher reductions. Less heat would therefore be generated, lowering the cooling load of the facility. Other benefits of a DC power delivery system are also possible. For example, Baldwin Technologies, which does system design, has promoted benefits of a DC power delivery system for data centers. These estimated benefits are based on vendor claims, rated performance of components, as well as improvements that Baldwin anticipates will derive from its own DC power delivery system design. The benefits and estimated performance improvements include the following:

- A lower number of components are needed, leading to lower maintenance costs and greater reliability.

- DC power distribution delivery is modular and flexible, so systems can grow with load requirements.

- Busways with double end-feed features allow for redundant DC sources at critical loads.

- No down-stream static or transfer switches are required, and voltage-matched DC systems can inherently be coupled together.

- DC distribution eliminates harmonics.

- Grounding is simplified.

- Management software and controls are available.

- DC distribution eliminates power factor concern.

- Server reliability may be increased by as much as 27%.

Baldwin's DC power system is being demonstrated at the Pentadyne Power facility in Chatsworth, California, which employs off-the-shelf equipment available from several manufacturers, including:

- **Rectifiers** that convert utility- or generator-supplied AC power to DC (500 VDC)

- **Energy storage**, in this case not batteries, but rather a flywheel-based system that can provide power to a 500 VDC bus if AC sources are lost

- **Equipment racks with DC distribution** entailing connectors that enable feeding power from two separate 500 VDC sources for redundancy

- **DC to DC converters** for conversion of 500 VDC power to low-voltage DC (e.g., 48V, 24V, 5V, etc.) as required by server equipment

YOUR FUTURE NEIGHBORHOOD

Adding DC power delivery systems to our homes, office buildings, or commercial facilities offers the potential for improvements in energy-delivery efficiency, reliability, power quality, and cost of operation as compared to traditional power systems. DC power distribution systems may also help overcome constraints in the development of new transmission capacity that are beginning to impact the power industry.

What might a future with DC power delivery look like? A number of options are available. One includes stand-alone systems that can operate full time as off-the-grid "islands," independent of the bulk power supply system. Hybrid buildings are also possible, with utility-supplied power as well as building-based generators such as a solar array, fuel cell, energy storage device, or even a hybrid automobile.

DC systems can operate selected loads or critical subsystems, such as computers and lights. Or a DC charging "rail" such as the kitchen countertop shown in Figure 5-1, can charge a host of portable appliances.

In fact, equipment throughout the entire house could be powered by DC, as shown in Figure 5-2.

POTENTIAL FUTURE WORK AND RESEARCH

Technology advances suggest that there are significant opportunities for certain DC-based applications, and promising benefits in terms of energy savings and increased reliability. But many obstacles must

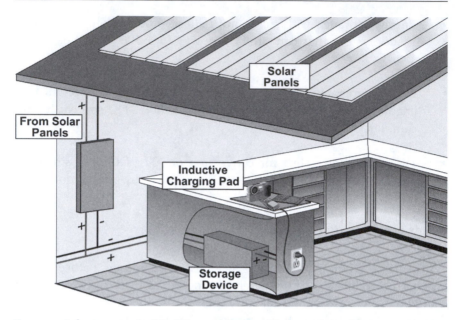

Tomorrow's homes may be blissfully cord free, enabling people to charge portable electronics using an inductive charging pad fed by rooftop solar cells

Figure 5-1. A DC-powered inductive charging system (www.galvinpower.org)

be overcome. Additional research, development and demonstration are needed to make DC systems viable. Below, we discuss some of the barriers and research needs presented by DC power delivery systems.

The Business Case for DC Power Delivery is Not Yet Clear—Will potential operating cost savings be sufficient to warrant initial capital investment for early adopters? For what applications? To what extent will DC play into new power delivery infrastructure investments? How, for example, can DC power systems enable use of plug-in hybrid vehicles, which may become tomorrow's mobile "mini" power plants? Systems that will accommodate efficient, safe, and reliable power delivery between such vehicles and either energy sources or loads are needed. Whether DC power systems are a practical option must be assessed.

Most Equipment is Not Yet Plug Ready; Demonstrations with Manufacturers are Called For—Even though electronic devices ultimately operate on DC, they have been designed with internal conversion systems to change AC to DC, and do not typically have ports for DC power delivery. Although some specific products are available to

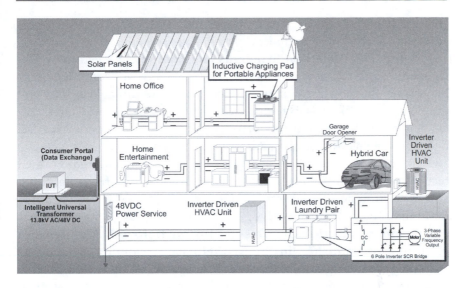

From the kitchen inductive charger to the PC to the air conditioner, appliances throughout the house could be DC powered.

Figure 5-2. A possible DC power system for tomorrow's home
(www.galvinpower.org)

accept DC power—such as DC fluorescent lighting ballasts, or server rack distribution systems—for most loads, AC 60-Hz power still must be supplied. Since the electronics market is highly competitive and has relatively low profit margins, a compelling business case is necessary before product designers and manufacturers will alter their products and add DC power ports—or make other changes to their equipment. To document potential and expand markets, additional demonstrations are needed with equipment that holds promise for use with DC power delivery, such as variable frequency drives.

For Data Center Applications, More Field Testing and Perfor- mance Measurements are Required—Several manufacturers have developed components that enable DC power delivery in data centers, including rectifiers, storage systems, DC to DC converters, and rack dis- tribution systems. However, the benefits of DC power delivery, such as energy efficiency, have only been estimated, based on vendor claims and rated performance of various components. Measured data on potential energy savings, as well as other performance metrics such as power reliability and power quality, the lifetime of converters, maintenance needs, and other factors are required.

Safety and Protection Standards and Equipment Need to be Developed—Since DC power does not cycle to a current "zero" 120 times per second like 60 Hz AC current does, it is more difficult to interrupt the flow of DC power. Therefore, DC power switches and interrupters employing semiconductors or other technology are needed for DC delivery systems. Also to be addressed are when and where solid-state switches need to be applied, and when an air gap is required. Further, techniques for controlling transients, such as spikes from lightning strikes, require additional investigation and testing—as does research for grounding and balancing DC.

Standard Practices for Design, Installation, and Maintenance Need to be Established in the Marketplace—Adoption of any new technology or design procedures can represent a significant hurdle. Designers, technicians, installers, retailers, buyers, and users want to mitigate risk and cost, which requires investment in product development, system integration, professional training—and time.

CONCLUSION

As the smart grid evolves, it may be appropriate to rethink the wider use of DC power distribution in buildings.

References
DC Power Production, Delivery and Utilization: An EPRI White Paper, The Electric Power Research Institute, June 2006.
Galvin Electricity Initiative: Transforming Electric Service Reliability and Value for the 21st Century, 2005, www.galvinpower.org.

Chapter 6

The IntelliGrid^{SM} Architecture
For the Smart Grid

The challenge before the energy industry remains formidable, for, as stated in the July 2001 issue of *Wired* magazine, "the current power infrastructure is as incompatible with the future as horse trails were to automobiles." But with an aggressive, public/private, coordinated effort the present power delivery system and market structure can be enhanced and augmented to meet the challenge it faces—evolving into an IntelliGrid^{SM}.

The 2003 blackout in the Northeast reminds us that electricity is indeed essential to our well-being. And it highlights one of the most fundamental of electric functions: getting electricity from the point of generation to the point of use. Power delivery has been part of the utility industry for so long that it is hard to imagine that this process has not already been optimized. However, the power delivery function is changing and growing more complex with the exciting requirements of the digital economy, the onset of competitive power markets, the implementation of modern and self-generation, and the saturation of existing transmission and distribution capacity. Without accelerated investment and careful policy analysis, the vulnerabilities already present in today's power system will continue to degrade.

Simply stated, today's electricity infrastructure is inadequate to meet rising consumer needs and expectations.

INTRODUCTION

The nation's power delivery system is being stressed in new ways for which it was not designed. For example, while there may have been specific operational, maintenance and performance issues that contributed to the August 14, 2003 outage, a number of improvements to the system could minimize the potential threat and severity of any future outages.

EPRI's study of the outage was performed during a three-week period immediately following and identified several areas that need further, rigorous investigation, including the effects of:

- **Reactive power reserves in the region.** Reactive power is the additional power required for maintaining voltage stability when serving certain kinds of energy consuming devices and appliances, such as motors, air conditioning, and fluorescent lights.

- **Power flow patterns over the entire region**, coupled with issues of coordination, control and communications of power system activities on a regional basis.

- **New power flows resulting from changing geographic patterns of consumer demand and the installation of new power plants.**

LAUNCHING THE INTELLIGRID^SM

As previously stated, today's electricity infrastructure is inadequate to meet rising consumer needs and expectations. A sharp decline in critical infrastructure investment over the last decade has already left portions of the electric power system vulnerable to poor power quality service interruptions and market dislocations. Substantial system upgrades are needed just to bring service back to the level of reliability and quality already required and expected by consumers, and to allow markets to function efficiently so that consumers can realize the promised benefit of industry restructuring.

To assure that the science and technology would be available to address the infrastructure needs, in 2001 the Electric Power Research Institute (EPRI) and the Electricity Innovation Institute (E2I) initiated an ambitious program. It was called the Consortium for Electric Infrastructure to Support a Digital Society (CEIDS) and hoped to build public/private partnerships to meet the energy needs of tomorrow's society. This effort essentially launched the smart grid concept—albeit not using that label. It was the CEIDS effort that formed the foundation for most all smart grid efforts to follow.

The CEIDS consortium believed that the restructuring and rise of the digital economy have set electricity price, quality and reliability on a collision course. The main driving force behind efforts to increase competition in both wholesale and retail power markets was the need to make inexpensive electricity more widely available—in particular, to reduce regional price inequities. Already, the effects of deregulation are being seen in the wholesale market, with both prices and price differentials declining rapidly. The effect on retail markets will come more slowly, but over the next 20 years, the average real price of electricity in the U.S. is expected to fall by 10% for residential customers and 14% for industrial customers. At the same time, however, industry restructuring has not yet provided adequate financial incentives for utilities to make the investments necessary to maintain—much less improve—power delivery quality and reliability.

CEIDS believes that meeting the energy requirements of society will require applying a combination of advanced technologies—from generating devices (e.g., conventional power plants, fuel cells, microturbines) to interface devices to end-use equipment and circuit boards. Simply "gold plating" the present delivery system would not be a feasible way to provide the level of security, quality, reliability and availability required. Neither will the ultimate customers themselves find traditional utility solutions satisfactory or optimal in supplying the ever-increasing reliability and quality of electric power they demand.

In addition, new technology is needed if society is to leverage the ever-expanding opportunities of communications and electric utilities' natural connectivity to consumers to revolutionize both the role of a rapidly changing industry and the way consumers may be connected to electricity markets of the future. CEIDS believed it could enable such a transformation and ushered the direction for building future infrastructure needed.

CEIDS initiated the creation of new levels of social expectations, business savvy and technical excellence by attracting players from the electric utility industry, manufacturers and end users as well as federal and state agencies. In order to achieve this, CEIDS was guided by the following key principles:

- *Vision*: To develop the science and technology that will ensure an adequate supply of high-quality, reliable electricity to meet the energy needs of the digital society.

- *Mission*: CEIDS provides the science and technology that will power a digital economy and integrate energy users and markets through a unique collaboration of public, private and governmental stakeholders.

CEIDS later morphed into the IntelliGrid^{SM} and the Electricity Innovation Institute was absorbed into EPRI. The consortium survives and prospers more than ever with over 50 collaborative partners.

THE INTELLIGRID^{SM} TODAY

The IntelliGrid^{SM} remains on course to provide the architecture for the smart grid by addressing five functionalities in the power system of today. These functionalities are consistent with those outlined in Chapter 1, and include the following:

Visualizing the Power System in Real Time
This attribute would deploy advanced sensors more broadly throughout the system on all critical components. These sensors would be integrated with a real-time communications system through an Integrated Electric and Simulation and Modeling computational ability and presented in a visual form in order for system operators to respond and administer.

Increasing System Capacity
This functionality embodies a generally straightforward effort to build or reinforce capacity particularly in the high-voltage system. This would include building more transmission circuits, bringing substations and lines up to NERC N-1 criteria, making improvements on data infrastructure, upgrading control centers, and updating protection schemes and relays.

Relieving Bottlenecks
This functionality allows the U.S. to eliminate many/most of the bottlenecks that currently limit a truly functional wholesale market and to assure system stability. In addition to increasing capacity, as described above, this functionality includes increasing power flow, enhanced voltage support, providing and allowing the operation of the electrical

system on a dynamic basis. This functionality would also require technology deployment to manage fault currents.

Enabling a Self-healing Grid

Once the functionalities discussed above are in place, then it is possible to consider controlling the system in real time. To enable this functionality will require wide-scale deployment of power electronic devices such as power electronic circuit breakers and flexible AC transmission technologies. These technologies will then provide the integration with an advanced control architecture to enable a self-healing system.

Enabling (Enhanced) Connectivity to Consumers

The functionalities described above assume the integration of a communication system throughout much of the power system. Once that system is present, connectivity to the ultimate consumers can be enhanced with communications. This enhancement will allow three new areas of functionality: one which relates directly to electricity services (e.g., added billing information of real-time pricing); one which involves services related to electricity (e.g., home security or appliance monitoring); and the third involves what are more generally thought of as communications services (e.g., data services).

The IntelliGrid^{SM} architecture includes a bold new concept called the EnergyPort^{SM} (see Chapter 8). The EnergyPort^{SM} is the consumer gateway now constrained by the meter, allowing price signals, decisions, communications, and network intelligence to flow back and forth through a seamless two-way portal. The EnergyPort^{SM} is the linchpin technology that leads to a fully functioning retail electricity marketplace with consumers responding (through microprocessor agents) to price signals. Specific capabilities of the EnergyPort^{SM} can include the following:

- Pricing and billing processes that would support real-time pricing

- Value-added services such as billing inquiries, service calls, outage and emergency services, power quality monitoring, and diagnostics

- Improved building and appliance standards

- Consumer energy management through sophisticated on-site energy management systems

- Easy "plug and play" connection of distributed energy resources

- Improved real-time system operations including dispatch, demand response, and loss identification

- Improved short-term load forecasting

- Improved long-term planning

A SMART GRID VISION BASED ON
THE INTELLIGRID^{SM} ARCHITECTURE

The IntelliGrid^{SM} will enable achievement of the following goals:

- Physical and information assets that are protected from man-made and natural threats, and a power delivery infrastructure that can be quickly restored in the event of attack or a disruption: A "self-healing grid."

- Extremely reliable delivery of the high-quality, "digital-grade" power needed by a growing number of critical electricity end uses.

- Availability of a wide range of "always-on, price-smart" electricity-related consumer and business services, including low-cost, high-value energy services that stimulate the economy and offer consumers greater control over energy usage and expenses.

- Minimized environmental and societal impact by improving use of the existing infrastructure; promoting development, implementation, and use of energy-efficient equipment and systems; and stimulating the development, implementation, and use of clean distributed energy resources and efficient combined heat and power technologies.

- Improve productivity growth rates, increased economic growth rates, and decreased electricity intensity (ratio of electricity use to gross domestic product, GDP).

BARRIERS TO ACHIEVING THIS VISION

To achieve this vision of the power delivery system and electricity markets, accelerated public/private research, design and development (RD&D), investment and careful policy analysis are needed to overcome the following barriers and vulnerabilities:

- The existing power delivery infrastructure is vulnerable to human error, natural disasters, and intentional physical and cyber attack.

- Investment in expansion and maintenance of this infrastructure is lagging, while electricity demand grows and will continue to grow.

- This infrastructure is not being expanded or enhanced to meet the demands of wholesale competition in the electric power industry, and does not facilitate connectivity between consumers and markets.

- Under continued stress, the present infrastructure cannot support levels of power, security, quality, reliability and availability (SQRA) needed for economic prosperity.

- The infrastructure does not adequately accommodate emerging beneficial technologies including distributed energy resources and energy storage, nor does it facilitate enormous business opportunities in retail electricity/information services.

- The present electric power delivery infrastructure was not designed to meet, and is unable to meet, the needs of a digital society—a society that relies on microprocessor-based devices in home, offices, commercial buildings, industrial facilities and vehicles.

Communication Architecture: The Foundation of the IntelliGrid^{SM}

To realize the vision of the IntelliGrid^{SM}, standardized communications architecture must first be developed and overlaid on today's power delivery system. This "Integrated Energy and Communications System Architecture" (IECSA) will be an open standards-based systems architecture for a data communications and distributed computing infrastructure. Several technical elements will constitute this infrastructure including, but not limited to, data networking, communications over a wide variety of physical media, and embedded computing technologies. IECSA will enable the automated monitoring and control of power delivery systems in real time, support deployment of technologies that increase the control and capacity of power delivery systems, enhance the performance of end-use digital devices that consumers employ, and enable consumer connectivity, thereby revolutionizing the value of consumer services. Note that the IntelliGrid^{SM} Architecture is free to anyone and can be downloaded from EPRI's web site at http://intelligrid.epri.com/.

These were the initial steps in developing the IntelliGrid^{SM}: to define clearly the scope of the requirements of the power system functions and to identify all the roles of the stakeholders. These were then included in an Integrated Energy and Communications System Architecture. There are many power system applications and a large number of potential stakeholders who already participate in power system operations. In the future, more stakeholders, such as customers responding to real-time process, DR owners selling energy and ancillary services into the electricity marketplace, and consumers demanding high quality will actively participate in power system operations. At the same time, new and expanded applications will be needed to respond to the increased pressures for managing power system reliability as market forces push the system to its limits. Power system security has also been recognized as crucial in the increasingly digital economy. The key is to identify and categorize all of these elements so that their requirements can be understood, their information needs can be identified, and eventually synergies among these information needs can be determined. One of the most powerful methodologies for identifying and organizing the pieces in this puzzle is to develop business models that identify a strawman set of entities and address the key interactions between these entities. These business models will establish a set of working relationships between industry entities in the present and the future, including intermediate steps from vertical operations to restructures operations. The business models will include, but not be limited to, the following:

- **Market operations**, including energy transactions, power system scheduling, congestion management, emergency power system management, metering, settlements and auditing.

- **Transmission operations**, including optimal operations under normal conditions, prevention of harmful contingencies, short-term operations planning, emergency control operations, transmission maintenance operations, and support of distribution system operations.

- **Distribution operations**, including coordinate volt/var control, automated distribution operation analysis, fault location/isolation, power restoration, feeder reconfiguration, DR management, and outage scheduling and data maintenance.

- **Customer services**, including automated meter reading (ARM), time-of-use and real-time pricing, meter management, aggregation for market participation, power quality monitoring, outage management, and in-building services and services using communications with end-use loads within customer facilities.

- **Generation at the transmission level**, including automatic generation control, generation maintenance scheduling, and coordination of wind farms.

- **Distributed resources at the distribution level**, including participation of DR in market operations, DR monitoring and control by non-utility stakeholders, microgrid management and DR maintenance management. These business models will be analyzed and used to define the initial process boundaries for subsequent tasks. Business process diagrams will be used to illustrate the more complex interactions.

The scope of IECSA architecture encompasses the power system from the generator to the end-use load. In other words, the IECSA architecture extends as far as the electric energy extends to do useful work. This means the IECSA architecture includes the distributed computing environments in in-building environments, as well as interaction with the foreseen operation of intelligent end-use subsystems and loads within the customer's facility.

The business models include the issues surrounding RTO/ISO operations and the seams issues between entities serving restructured as well as vertical markets. This includes business operations that span across domains such as customer participation in ancillary service functions as well as self-healing grid functionality. Standard documentation languages such as unified modeling language (UML), high-level architecture (HLA), and others are incorporated as appropriate. The intent here is to make the resulting information useful for the various stakeholder groups. Business entities include vertical utilities, as well as business entities anticipated to participate in a fully restructured electric and energy service industry. The business models also included key communications business models that could provide either common services or private network infrastructures to support the power industry through the IECSA architecture. The business models for communications include generic communications business functions that may have roles in the ultimate implementation of the IECSA including, but not limited to, common carrier services and the provision of access network technologies to customers. These business models would address the application of new communications technologies that exhibit self-healing capabilities similar to that proposed for the power system

Fast Simulation and Modeling

Once the IECSA begins to be deployed, then the industry's computational capability must be enhanced. In order to enable the functionality of the power delivery system, a capability which allows fast simulation and modeling (FSM) will need to evolve in order to assure the mathematical underpinning and look-ahead capability for a self-healing grid (SHG)—one capable of automatically anticipating and responding to power system disturbances, while continually optimizing its own performance. Creating a SHG will require the judicious use of numerous intelligent sensors and communication devices that will be integrated with power system control through a new architecture that is being developed by CEIDS. This new architecture will provide a framework for fundamentally changing system functionality as required in the future. The FSM project will augment these capabilities in three ways:

- Provide faster-than-real-time, look-ahead simulations and thus be able to avoid previously unforeseen disturbances.

- Perform what-if analyses for large-region power systems from both operations and planning points of view.

- Integrate market, policy and risk analyses into system models and quantify their effects on system security and reliability.

The next step in creating a SHG will involve addition of intelligent network agents (INAs) that gather and communicate system data, make decisions about local control functions (such as switching a protective relay), and coordinate such decisions with overall system requirements. Because most control agents on today's power systems are simply programmed to respond to disturbances in pre-determined ways—for example, by switching off a relay at a certain voltage—the activity may actually make an incipient problem worse and contribute to a cascading outage. The new simulation tools developed in the FSM project will help prevent such cascading effects by creating better system models that use real-time data coming from INAs over a wide area and, in turn, coordinate the control functions of the INAs for overall system benefit, instead of the benefit of one circuit or one device. As discussed later, having such improved modeling capability will also enable planners to better determine the effects of various market designs or policy changes on power system reliability.

To reach these goals, the FSM project will focus on the following three areas:

- **Multi-resolution Modeling.** New modeling capabilities will be developed that provide much faster analysis of system conditions and offer operators the ability to "zoom" in or out to visualize parts of a system with lower or higher degrees of resolution. This off-line modeling activity will be the focus of work during the first two years of the project.

- **Modeling of Market and Policy Impacts on Reliability.** The recent western states power "crisis" dramatically illustrated how untested policies and market dynamics can affect power system reliability. The new modeling capabilities being developed in this project will allow planners and policymakers to simulate the effects of their activities before actually putting them into practice. This effort will be conducted in parallel with the previous project, beginning in the first year, and then integrated with the multi-resolution system models.

- **Validation of Integrated Models with Real-Time Data.** Once the new, integrated models have been thoroughly tested off-line, they will be validated and enhanced using real-time data from major power systems. This work will begin in the third year and continue through the end of the project. Full integration with on-line network control functions and INAs will be left to a follow-on project.

Unified integration architecture is the key enabler to successfully and inexpensively deploying advanced functions. This architecture must be robust enough to meet the numerous disparate requirements for power system operations and be flexible enough to handle changing needs. The focus of this project is to identify and propose potential solutions for enterprise and industry-wide architectural issues that will arise from the high levels of integration and management foreseen by this project. The concepts and functions defined in the IECSA will support the development and deployment of distributed applications that reach to and across a great number of applications and stakeholders. The IECSA is required to integrate customer interaction, power system monitoring and control, energy trading, and business systems. It will reach across customers, feeders, substation, control centers, and energy traders. The IECSA must also provide a scalable and cohesive way to access resources across the wide spectrum of applications while at the same time providing the means to filter out unwanted data.

Open Communication Architecture for
Distributed Energy Resources in Advanced Automation

A subset of the work on an Integrated Energy and Communications System Architecture is the development of an open communication architecture for distributed energy resources (DER) in advanced distribution automation (ADA) or DER/ADA architecture.

The DER/ADA Architecture Project will develop the object models for integration of specific DER types into the open communication architecture that is being developed through the companion CEIDS project know as the Integrated Energy and Communication Systems Architecture (IECSA) Project. Whereas the IECSA Project is concerned with the broad requirements for the architecture, the DER/ADA Architecture

Project is focused on a very specific, but very important, piece of the whole—object models for DER devices.

It is important to note that the term DER has various definitions and is used ambiguously in different situations. The broadest use of the term includes distributed generation, storage, load management, combined heat and power, and other technologies involved in electricity supply, both in stand-alone (off-grid) applications and in applications involving interconnection with power distribution systems. Another term, distributed resources (DR), is similarly used in an ambiguous manner.

Eventually, object models will also be needed for other DER types besides distributed generation and storage. Hence, for the remainder of this chapter, DER is used for brevity to denote distributed generation and storage.

Key benefits that will be derived from the object model development in this project and from the broader open architecture development in the IECSA Project include:

- Increasing the functionality and value of DER in distribution system operations, which benefits both the utility and the consumer of electricity.

- Growing the market for DER equipment vendors.

- Providing a large market to communication and control equipment providers for sale of their products to help build the infrastructure.

ENABLING TECHNOLOGIES

In addition to the IECSA foundation, EPRI has developed the following list of critical enabling technologies that are needed to move toward realizing IntelliGridSM:

- Automation: the heart of the IntelliGridSM
- Distributed energy resources and storage development and integration
- Power electronics-based controllers
- Power market tools

- Technology innovation in electricity use

- The Consumer Portal

These technologies are synergistic (i.e., they support realization of multiple aspects of the vision). Aspects of some of these enabling technologies are under development today. Each of these technologies calls for either continued emphasis or initiation of efforts soon in order to meet the energy needs of society in the next 20 years and beyond.

Automation: The Heart of the IntelliGrid^SM

Automation will playa key role in providing high levels of power SQRA throughout the electricity value chain of the near future. To a consumer, automation may mean receiving hourly electricity price signals, which can automatically adjust home thermostat settings via a smart consumer portal. To a distribution system operator, automation may mean automatic "islanding" of a distribution feeder with local distributed energy resources in an emergency. To a power system operator, automation means a self-healing, self-optimizing smart power delivery system that automatically anticipates and quickly responds to disturbances to minimize their impact, minimizing or eliminating power disruptions altogether. This smart power delivery system will also enable a revolution in consumer services via sophisticated retail markets. Through a two-way consumer portal that replaces today's electric meter, consumers will tie into this smart power delivery system. This will allow price signals, decisions, communications, and network intelligence to efficiently flow back and forth between consumer and service provider in real time. The resulting fully functioning retail marketplace will offer consumers a wide range of services, including premium power options, real-time power quality monitoring, home automation services, and much more.

Distributed Energy Resources and
Storage Development & Integration

Small power generation and storage devices distributed throughout—and seamlessly integrated with—the power delivery system ("distributed energy resources") and bulk storage technologies offer potential solutions to several challenges that the electric power industry currently faces. These challenges include the needs to strengthen the power delivery infrastructure, provide high-quality power, facilitate provision of a range of services to consumers, and provide consumers lower-cost, higher-SQRA power. However, various impediments stand in the way of widespread realization of these benefits. A key challenge for distributed generation and storage technologies, for example, is to develop ways of seamlessly integrating these devices into the power delivery system, and then dispatching them so that they can contribute to overall reliability and power quality. Both distributed storage and bulk storage technologies address the inefficiencies inherent in the fact that, unlike other commodities, almost all electricity today must be used at the instant it is produced.

Power Electronics-based Controllers

Power electronics-based controllers, based on solid-state devices, offer control of the power delivery system with the speed and accuracy of a microprocessor, but at a power level 500 million times higher. These controllers allow utilities and power system operators to direct power along specific corridors—meaning that the physical flow of power can be aligned with commercial power transactions. In many instances, power electronics-based controllers can increase power transfer capacity by up to 50% and, by eliminating power bottlenecks, extend the market reach of competitive power generation. On distribution systems, converter-based power electronics technology can also help solve power quality problems such as voltage sags, voltage flicker, and harmonics.

Power Market Tools

To accommodate changes in retail power markets worldwide, market-based mechanisms are needed that offer incentives to market participants in ways that benefit all stakeholders, facilitate efficient planning for expansion of the power delivery infrastructure, effectively allocate risk, and connect consumers to markets. For example, service providers need a new methodology for the design of retail service programs for electricity consumers. At the same time, consumers need help devising ways they can participate profitably in markets by providing dispatchable or curtailable electric loads, especially by providing reserves. And market participants critically need new ways to manage financial risk. To enable the efficient operation of both wholesale and retail markets, rapid, open access to data is essential. Hence, development of data and communications standards for emerging markets is needed. Further, to test the viability of various wholesale and retail power market design options before they are put into practice, power market simulation tools are needed to help stakeholders establish equitable power markets.

Technology Innovation in Electricity Use

Technology innovation in electricity use is a cornerstone of global economic progress. In the U.S., for example, the growth in GDP over the past 50 years has been accompanied by improvements in energy intensity and labor productivity. Improved energy-use efficiencies also provide environmental benefits. Development and adoption of technologies in the following areas are needed:

- Industrial electrotechnologies and motor systems
- Improvement in indoor air quality
- Advanced lighting
- Automated electronic equipment recycling processes

In addition, widespread use of electric transportation solutions—including hybrid and fuel cell vehicles—will reduce petroleum consumption, reduce the U.S. trade deficit, enhance U.S. GDP, reduce emissions, and provide other benefits.

The Consumer Portal

Once communications and electricity infrastructures are integrated, realizing the ability to connect electricity consumers more fully with electronic communications will depend on evolving a consumer portal to function as a "front door" to consumers and their intelligent equipment. The portal would sit between consumers "in-building" communications network and wide area "access" networks. The portal would enable two-way, secure and managed communications between consumers equipment and energy service and/or communications entities. It would perform the work closely related to "routers" and "gateways" with added management features to enable energy industry networked applications including expanded choice; real-time pricing; detailed billing and consumption information; wide area communications and distributed computing. This could include data management, and network access based on consumer systems consisting of in-building networks and networked equipment which integrate building energy management, distributed energy resources, and demand response capability with utility distribution operations.

CONCLUSION

The participation of energy companies, universities, government and regulatory agencies, technology companies, associations, public advocacy organizations, and other interested parties throughout the world need to contribute to refining the vision and evolving the needed technology. Only through collaboration can the resources and commitment be marshaled to enable the IntelliGrid^{SM}.

Chapter 7

The Smart Grid—
Enabling Demand Response—
The Dynamic
Energy Systems Concept

Dynamic energy systems provide the infrastructure to use the smart grid to enable demand response through dynamic energy management systems. Dynamic energy management is an innovative approach to managing load at the demand-side. It incorporates the conventional energy use management principles represented in demand-side management, demand response, and distributed energy resource programs and merges them in an integrated framework that simultaneously addresses permanent energy savings, permanent demand reductions, and temporary peak load reductions. This is accomplished through an integrated system comprised of smart end-use devices and distributed energy resources with highly advanced controls and communications capabilities that enable dynamic management of the system as a whole. This simultaneous implementation of measures sets this approach apart from conventional energy use management and eliminates any inherent inefficiencies that may otherwise arise from a piecemeal deployment strategy. It offers a "no-regrets" alternative to program implementers.

Dynamic energy management consists of four main components:

1. Smart energy efficient end-use devices
2. Smart distributed energy resources
3. Advanced whole-building control systems
4. Integrated communications architecture

These components act as building blocks of the dynamic energy management concept. The components build upon each other and interact with one another to contribute to an infrastructure that is dynamic, fully integrated, highly energy-efficient, automated and capable of learning. The result is an infrastructure comprised of individual elements that are capable of working in unison to optimize operation of the integrated system based on consumer requirements, utility constraints, available incentives and other variables such as weather and building occupancy. The following bullet points summarize the predominant characteristics of each of these three components. The components and how they potentially interplay will be covered in greater detail later in this white paper.

SMART ENERGY EFFICIENT END-USE DEVICES

• Appliances, lighting, space conditioning, and industrial process equipment with the highest energy efficiencies technically and economically feasible.

• Thermal energy storage systems that allow for load shaping.

• Intelligent end-use devices equipped with embedded features allowing for two-way communications and automated control.

• Devices that represent an evolution from static devices to dynamic devices with advancements in distributed intelligence; one example is a high-efficiency, Internet protocol (IP) addressable appliance that can be controlled by external signals from the utility, end-user or other authorized entity.

SMART DISTRIBUTED ENERGY RESOURCES

• On-site generation devices such as photovoltaics, diesel engines, micro-turbines and fuel cells that provide power alone or in conjunction with the grid.

• On-site electric energy storage devices such as batteries and fly wheels.

- Devices that are dynamically controlled to supply baseload, peak shaving, temporary demand reductions or power quality.

- Devices that are dynamically controlled such that excess power is sold back to the grid.

ADVANCED WHOLE-BUILDING CONTROL SYSTEMS

- Control systems that optimize the performance of end-use devices and distributed energy resources based on operational requirements, user preferences and external signals from the utility, end-user or other authorized entity.

- Controls that ensure end-use devices only operate as needed; examples include automatic dimming of lights when daylighting conditions allow or reducing outdoor ventilation during periods of low occupancy.

- Controls that allow for two-way communications; for example, they can send data (such as carbon dioxide concentration in a particular room) to an external source and they can accept commands from an external source (such as management of space conditioning system operation based on forecasted outside air temperature).

- Local, individual controls that are mutually compatible with a whole-building control system; for example, security, lighting, space conditioning, appliances, distributed energy resources, etc., can all be controlled by a central unit.

- Controls that have the ability to learn from past experience and apply that knowledge to future events.

INTEGRATED COMMUNICATIONS ARCHITECTURE

- Allow automated control of end-use devices and distributed energy resources in response to various signals such as pricing or emergency demand reduction signals from the utility; day-ahead weather forecasts; other external alerts (e.g., a signal could be sent to shut down the outdoor ventilation systems in the building in the event of a chemical attack in the area); and end-user signals (e.g., a facility manager could shut down the building systems from an off-site location during an unscheduled building closure.

• Allow the end-use devices, distributed energy resources and/or control systems to send operational data to external parties (e.g., advanced meters that communicate directly with utilities).

• Communications systems that have an open architecture to enable interoperability and communications among devices.

Figure 7-1 shows an example of the dynamic energy management infrastructure applied to a generic building. In this example, there are two-way communications via the Internet as well as via the power line. The building is equipped with smart energy-efficient end-use devices, an energy management system (EMS), automated controls with data management capabilities, and distributed energy resources such as solar photovoltaics, wind turbines and other on-site generation and storage systems. Thus, energy-efficient devices, controls and demand response strategies are coupled with on-site energy sources to serve as an additional energy "resource" for the utility. Not only do all of these elements contribute to the utility's supply-side by reducing building demand, the distributed energy resources can also feed excess power back to the grid.

A dynamic energy management system is likely to have a much larger impact on a building's electricity consumption and demand than just implementing energy efficiency and/or demand response on their own. Example 7-1 illustrates this point for an office building in the U.S.

One of the key enabling characteristics of the Dynamic Energy Management framework would be a smart grid. The following subsection describes a recent assessment that quantifies the energy savings potential associated with a smart grid.

ENERGY MANAGEMENT TODAY

Current practice in the implementation phase of energy use management consists of several elements used alone or in combination to

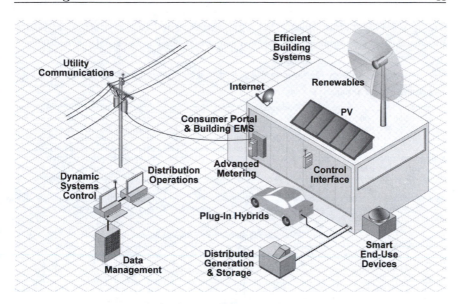

Figure 7-1. The Dynamic Energy Management Infrastructure Applied to a Generic Building

effect a change in energy use characteristics at a given site. In general, the elements can be divided into seven main categories:

1. Energy audits and/or reviews of historical energy use characteristics to identify problem areas.

2. Improvements to the operation and maintenance of existing end-use devices and processes to reduce energy use, demand and/or materials. This includes housekeeping and maintenance measures, heat recovery, energy cascading, material recovery/waste reduction, etc.

3. Replacement or retrofit of existing end-use devices or processes with energy-efficient devices to reduce energy use, demand and/or materials as well as to improve productivity. This may also include fuel switching (e.g., from thermal processes to electrotechnologies).

4. Load shaping strategies such as thermal energy storage which shifts load to off-peak periods.

Example 7-1. Dynamic Energy Management Applied to a Hypothetical Office Building. Source: *Dynamic Energy Management*, **EPRI, Palo Alto, CA: 2007.**

Dynamic Energy Management Applied to a Hypothetical Office Building

Modeling Scenarios and Assumptions

In this example, a DOE-2 model of a new office building in Atlanta, Georgia is simulated under several different scenarios in order to evaluate the building's electricity consumption and demand under each scenario. The scenarios are: (1) the Base Case Scenario that represents 'business as usual,' (2) the Energy Efficiency Scenario that represents the installation and implementation of energy efficiency measures that are typically promoted in utility energy efficiency programs, (3) the Demand Response (DR) Scenario that represents the implementation of manual demand response strategies during extremely warm summer days, (4) the Combination Scenario that represents the simultaneous implementation of both energy efficiency and demand response as defined by #2 and #3 above, and (5) the Dynamic Energy Management Scenario that represents the implementation of Dynamic Energy Management technologies and strategies. Only measures and strategies that affect the lighting and cooling end-uses are included in the modeling of these scenarios. (For more information on the modeling procedure and parameters used in this example, please refer to *Dynamic Energy Management*, EPRI, Palo Alto, CA: 2007). The table below describes the strategies that were modeled in the base case and each of the scenarios.

Specific Strategies Modeled Under Different Scenarios

Measure/ Strategy	Energy Efficiency Scenario	Demand Response Scenario	Combination of Energy-Efficiency and Demand Response Scenario	Dynamic Energy Management Scenario
1. Cooling	Involves installing a higher-efficiency chiller than standard practice.	Involves manually setting the cooling setpoint to 79 F during a DR event. According to the Auto DR pilot project conducted in PG&E's service territory, approximately 10% of an office building's total demand can be reduced by implementing this strategy.	Involves installing a higher-efficiency chiller than standard practice. Also involves manually setting the cooling setpoint to 79 F during a DR event.	Involves installing the highest-efficiency chiller available. Also involves manually setting the cooling setpoint to 79 F during 10 critical peak pricing days and/or DR events throughout the year.
2. Envelope Insulation	Involves installing a higher-efficiency ceiling and roof insulation than standard practice.		Involves installing a higher-efficiency ceiling and roof insulation than standard practice.	Involves installing a higher-efficiency ceiling and roof insulation than standard practice.
3. Windows/ Glazing	Involves installing Low-E glazing.		Involves installing Low-E glazing.	Involves installing Low-E glazing.

(The table is continued on the following page)

(Continued)

5. Installation of controls to turn end-use devices "on/off" or "up/down" as required or desired to reduce energy use and/or demand. This includes local controls and building energy management systems.

6. Demand response strategies to reduce peak demand temporarily.

7. Use of distributed energy resources to replace or reduce dependence on electricity from the grid.

Example 7-1. (Cont'd)

Measure/ Strategy	Energy Efficiency Scenario	Demand Response Scenario	Combination of Energy-Efficiency and Demand Response Scenario	Dynamic Energy Management Scenario
4. Lighting	Involves implementing typical lighting measures such as T-5 lamps, super T-8 lamps, compact fluorescent lamps, task lighting, occupancy sensors, and time clocks and controls.	Includes manually turning off lights completely in unused areas and manually turning off a portion of lights in other areas ("bi-level switching") during a DR event. Assumes that 15% of lighting power can be reduced by implementing these strategies.	Involves implementing typical lighting measures such as T-5 lamps, super T-8 lamps, compact fluorescent lamps, task lighting, occupancy sensors, and controls. Also involves manually turning off lights completely in unused areas and manually turning off a portion of lights in other areas ("bi-level switching") during a DR event.	Involves implementing typical lighting measures such as T-5 lamps, super T-8 lamps, compact fluorescent lamps, task lighting, dimmable ballasts, occupancy sensors, and time clocks. Additionally, advanced lighting controls are used to integrate daylighting into lamp dimming schemes. Also the advanced lighting controls will automate the process of dimming and/or turning off lights during 10 critical peak pricing and/or DR events throughout the year. Assumes that 30% of lighting power can be reduced by implementing these strategies.
5. Other	Includes optimizing set points of zone thermostats and cooling system operation schedules.		Includes optimizing set points of zone thermostats and cooling system operation schedules.	Includes optimizing set points of zone thermostats and cooling system operation schedules, automation of the chilled water and condenser water reset functions, and automation of pre-cooling the building in anticipation of warm days throughout the year. During 10 critical peak pricing and/or DR events during the year, the control system will also automatically implement the following strategies: increase thermostat set points (as described in #1 above), increase duct static pressure, increase chilled water temperature, and limit the distribution VSD fan speed. According to the Auto DR pilot project conducted in PG&E's service territory, approximately 15% of an office building's total demand can be reduced by implementing these strategies.

Results

Scenario	Base Case		Energy Efficiency/Demand Response/Dynamic Energy Management Implementation		Savings			
	kWh/yr	kW	kWh/yr	kW	kWh/yr	%	kW	%
Energy Efficiency Scenario	1,317,350	365 [1]	1,014,762	284 [1]	302,588	23%	81 [2]	22%
Demand Response Scenario	1,317,350	365 [1]	1,317,350 [3]	312 [1]	0 [3]	0%	53 [2]	15%
Energy Efficiency + Demand Response Scenario	1,317,350	365 [1]	1,014,762 [3]	243 [1]	302,588 [3]	23%	122 [2]	33%
Dynamic Energy Management Scenario	1,317,350	365 [1]	844,877	185 [1]	472,473	36%	180 [2]	49%

Notes:
1. This is the average electricity demand of the entire building during the three warmest weekdays of the summer (June 30 to July 2 of the typical meteorological year [TMY] climate data) between the hours of 2pm and 5pm.
2. This is the average electricity demand reduction of the entire building during the three warmest weekdays of the summer (June 30 to July 2 of the TMY climate data) between the hours of 2pm and 5pm.
3. Assumes that the energy savings due to demand response measures are negligible.

The modeling of the office building indicates that implementing energy-efficiency measures would result in electricity savings of 23% and a peak demand reduction of 22% relative to the base case. Implementing demand response strategies without implementing any energy efficiency measures would result in a peak demand reduction of 15% relative to the base case. (Energy savings due to demand response strategies are considered negligible because demand response actions are implemented during only a few emergency events throughout the year.) Using this same assumption, implementing a combination of energy efficiency and demand response in the same building would result in

(Continued)

Example 7-1. (Cont'd)

the same amount of electricity savings as the energy efficiency scenario (23%), but peak demand reduction would increase to 33% relative to the base case. The modeling results indicate that implementing Dynamic Energy Management would achieve a higher level of energy savings and demand reduction than implementing energy efficiency or demand response alone (or even the combination of the two). Implementing Dynamic Energy Management would result in an electricity savings of 36% and a peak demand reduction of 50% relative to the base case.

The important point here is that Dynamic Energy Management serves to fill in the gap and captures additional energy savings and demand reduction potential that would not otherwise be captured by implementing energy efficiency and demand response alone.

Source: *Dynamic Energy Management*, EPRI, Palo Alto, CA: 2007.

In some cases, the end-user takes the initiative to employ one or more of the elements listed above. Oftentimes, however, implementers of various energy use programs solicit participants. The elements are typically applied separately or in a piecemeal fashion, with the types of measures implemented being a strong function of the programs and incentives offered by implementers to program participants. Economic evaluations are also a key component to energy use management programs to quantify expected costs, savings, payback periods and returns on investment.

All of these elements fall within the framework of demand-side management in its broadest sense. However, typical practice compartmentalizes the elements into three main types of programs. Specifically, the first five elements are conventionally considered to be encompassed in *demand-side management* programs, while the last two elements are often considered separately and fall within *demand response* and *distributed energy resource* programs, respectively. The current practices within each of these three conventional categories are discussed next.

Demand-side Management

"Demand-side management is the planning, implementation and monitoring of those utility activities designed to influence customer use of electricity in ways that will produce desired changes in the utility's load shape, i.e., changes in the time pattern and magnitude of a utility's load. Utility programs falling under the umbrella of demand-side management include load management, new uses, strategic conservation, electrification, customer generation and adjustments in market share."

EPRI coined the term demand-side management in the early 1980s and continued to popularize the term through a series of more than 100 articles since that time including the five volume set *Demand-side Man-*

agement which is widely recognized as a definitive and practical source of information on the demand-side management process. During its peak of activity, annual demand-side management expenditures in the U.S. were measured in billions of dollars, energy savings were measured in billions of kWh, and peak load reductions were stated in thousands of MW. While activities nationally have slowed since then, demand-side management continues to influence the demand for electricity.

Demand-side management is even more encompassing than the above definition implies, because it includes the management of all forms of energy at the demand side, not just electricity. In addition, groups other than just electric utilities (including natural gas suppliers, government organizations, non-profit groups and private parties) implement demand-side management programs.

In general, demand-side management embraces the following critical aspects of energy planning:

* Demand-side management *will influence customer use.* Any program intended to influence the customer's use of energy is considered demand-side management.

* Demand-side management *must achieve selected objectives.* To constitute a desired load shape change, the program must further the achievement of selected objectives (i.e., it must result in reductions in average rates, improvements in customer satisfaction, achievement of reliability targets, etc.).

* Demand-side management *will be evaluated against non-demand-side management alternatives.* The concept also requires that selected demand-side management programs further these objectives to at least as great an extent as non-demand-side management alternatives, such as generating units, purchased power or supply-side storage devices. In other words, it requires that demand-side management alternatives be compared to supply-side alternatives. It is at this stage of evaluation that demand-side management becomes part of the integrated resource planning process.

* Demand-side management *identifies how customers will respond.* Demand-side management is pragmatically oriented. Normative programs ("we ought to do this") do not bring about the desired

change; positive efforts ("if we do this, that will happen") are required. Thus, demand-side management encompasses a process that identifies how customers will respond, not how they should respond.

- Demand-side management **value is influenced by load shape**. Finally, this definition of demand-side management focuses upon the load shape. This implies an evaluation process that examines the value of programs according to how they influence costs and benefits throughout the day, week, month and year.

Thus, the term demand-side management is extremely broad in its original intent, encompassing all actions that meet the critical aspects of energy planning listed above. In addition, it embodies all of the seven elements (listed previously) that are associated with current practice in energy use management. It can even be considered to encompass most of the essence of dynamic energy management. However, the two-way communications fundamental to the dynamic feature of dynamic energy management were not available during the inception of the demand-side management concept, and so they serve to update the demand-side management vision. Moreover, demand-side management does not inherently prescribe that the elements be implemented simultaneously as does dynamic energy management. Furthermore, despite its broad definition, demand-side management mainly results in the implementation of four main types of components in conventional implementation. These components include (1) energy-efficient end-use devices (which includes modification to existing devices and processes as well as new energy-efficient devices and processes); (2) additional equipment, systems and controls enabling load shaping (such as thermal energy storage devices); (3) standard control systems to turn end-use devices "on/off" or "up/down" as required or desired; and (4) the potential for communications between the end-user and an external party (however, this is generally not employed to a great extent).

Oftentimes, the components are implemented separately rather than simultaneously. For example, an energy-efficient lighting program may offer incentives for conversion from T-12 lamps and magnetic ballasts to T-8 lamps and electronic ballasts, but it may not offer incentives for lighting controls or other measures. In addition, a water-heating program may offer incentives specifically for conversion from gas to electric water heaters or to heat pump water heaters. Still other programs offer incentives for whole-building energy savings, regardless of how the savings are achieved. All of these measures positively affect both the energy companies and the customers to some extent. However, if a building-wide program incorporating a variety of measures coupled with a dynamic link between the end-use devices, controllers and energy suppliers were undertaken, the benefits would be optimized. This is where *dynamic* demand-side management, or the term EPRI has defined—*dynamic energy management*—comes to play.

Demand Response

Demand response (DR) refers to mechanisms to manage the demand from customers in response to supply conditions, for example, having electricity customers reduce their consumption at critical times or in response to market prices. There has been a recent upsurge in interest and activity in demand response, primarily due to the tight supply conditions in certain regions of the country that have created a need for resources that can be quickly deployed. Demand response can broadly be of two types—incentive-based demand response and time-based rates. Incentive-based demand response includes direct load control, interruptible/curtailable rates, demand bidding/buyback programs, emergency demand response programs, capacity market programs, and ancillary services market programs. Time-based rates include time-of-use rates, critical-peak pricing, and real-time pricing. Incentive-based demand response programs offer payments for customers to reduce their electricity usage during periods of system need or stress and are triggered either for reliability or economic reasons. A range of time-based rates is currently offered directly to retail customers with the objective of promoting customer demand response based on price signals. These two broad categories of demand response are highly interconnected, and the various programs under each category can be designed to achieve complementary goals.

According to a recent Federal Energy Regulatory Commission (FERC) study, nationally, the total potential demand response resource contribution from existing programs is estimated to be about 37,500 MW. The vast majority of this resource potential is associated with incentive-based demand response. This represented approximately 5% of the total U.S. projected electricity demand for summer 2006.

ROLE OF TECHNOLOGY IN DEMAND RESPONSE

Technology plays a key role in enabling demand response. The future growth of the demand response market capability depends on the cost, functionality and degree of process automation of technologies that enable demand response. Enabling technologies for demand response include:

•   Interval meters with two-way communications capability which allow customer utility bills to reflect their actual usage pattern and provide customers with continuous access to their energy use data.

•   Multiple, user-friendly, communication pathways to notify customers of real-time pricing conditions, potential generation shortages, as well as emergency load curtailment events.

•   Energy information tools that enable real-time or near-real-time access to interval load data, analyze load curtailment performance relative to baseline usage, and provide diagnostics to facility operators on potential loads to target for curtailment.

•   Demand reduction strategies that are optimized to meet differing high-price or electric system emergency scenarios.

•   Load controllers and building energy management control systems (EMCS) that are optimized for demand response, and which facilitate automation of load control strategies at the end-use level.

•   On-site generation equipment used either for emergency backup or to meet primary power needs of a facility.

Advancements in technologies regarding control systems, telecommunication and metering all increase the opportunity for end-users to monitor and adjust their electricity consumption in coordination with electricity market conditions. Additionally, distributed generation is an important source of supply when traditional supply sources become scarce. Developing these alternatives and incorporating them into the marketplace is increasingly becoming a reality and should provide increased demand-side responsiveness in the future. A service area that is growing rapidly is associated with automated meter reading (AMR) and advanced metering technologies. Customers are also investing in sophisticated energy information systems (EISs) and EMCSs—especially Internet-based controls for their facilities. Several vendors are selling proprietary hardware and software services for building energy management controls and systems. One way to drive the costs down for demand response, and to enhance the utilization of this program, is for utilities and customers to have common architecture for demand response-related activities, EIS/EMCS and utility-sponsored AMR programs. Integrated architectures require the implementation of common protocols used in today's Internet technology. Demand responsive control systems integrate the controls for the distributed (demand-responsive) energy system with electronic communication and metering technology to facilitate one-way or two-way communication between utility and customer equipment. These technologies are used to reduce energy use (by dimming lights, raising air-conditioning set points, etc.) in response to peak electricity demand emergencies and/or prices.

CURRENT LIMITATIONS AND
SCOPE FOR DYNAMIC ENERGY MANAGEMENT

Currently, the demand response enabling technologies have limitations in terms of system scaling and interoperation with other similar systems that impair their ability to be scaled up to serve the entire industry. Also, the individual demand response-enabling technology components discussed here are oftentimes implemented in a piecemeal fashion without integration of the different technology components. This results in demand response programs falling far short of the anticipated potential benefits associated with an integrated strategy to manage load.

Demand response functions are often applied to standard end-use devices, with local control systems and one-way or basic two-way communications. Utilities or other energy service providers have not yet implemented for the most part the full functionalities associated with the enabling technologies due to a number of technological, regulatory and economic barriers. Customers are often offered individual demand response programs instead of a single service offering comprised of different options to manage their electricity load, which ultimately results in a low level of the potential being realized. Furthermore, very rarely are energy service providers talking about integrating energy efficiency and demand response into a single offering. Virtually all energy efficiency programs, from market transformation programs (appliances and building codes) to immediate resource acquisition programs (rebates and performance contracting) help to lower system peaks, even if peak reduction is not the primary program goal. Customers often do not connect their participation in energy efficiency programs with demand response, because they do not understand that reducing their peak usage changes the system load profile and makes the electricity system more efficient. Energy efficiency can reduce load significantly, and the load reductions occur over many hours of the load shape and for many days of the year.

Distributed Energy Resources

In their most general sense, distributed energy resources include technologies for distributed generation (non-renewable and renewable), combined heat and power, energy storage, power quality, and even demand-side management and demand response. Since demand-side management and demand response have been treated separately herein, the current scope of distributed energy resources will include energy generation and storage technologies, including the generation of heat and power, and the storage of electricity. (Thermal energy storage was encompassed within demand-side management.) Distributed energy resources can be applied at the utility-scale where they feed into the distribution system, or they can be applied at the building level. The focus here is building-level distributed energy resources since they can be considered a demand-side energy alternative.

The principal purposes of distributed energy resources are:

1.   To supply stand-alone power and/or heat (e.g., for remote locations).

2.   To augment power from the grid (e.g., to minimize power purchases).

3.   To reduce transmission and distribution losses by placing power and energy sources closer to loads.

4.   To provide peak shaving or load leveling (e.g., to reduce peak demand costs and/or to enable participation in demand reduction programs).

5.   To guarantee power quality, reliability and security (e.g., for critical operations and processes).

6.   To reduce capital cost of transmission facility construction.

Some distributed energy resource technologies include:

- Solar photovoltaics

- Reciprocating engines

- Stirling engines

- Combustion turbines

- Microturbines

- Wind turbines

- Fuel cells (e.g., phosphoric acid, molten carbonate, polymer electrolyte membrane)

- Batteries (e.g., lead acid, nickel-cadmium, nickel-metal hydride, lithium ion, vanadium redox, zinc-bromine, sodium-sulfur, solid oxide)

- Superconducting magnetic energy storage

- Flywheel energy storage

- Ultracapacitors for storage

Some of these technologies are significantly more established and implemented than others. Some are still in the research and development stage. For example, diesel and gas reciprocating engines and gas turbines are well-established distributed generation technologies for large commercial and industrial buildings. The vast majority of these units are installed to serve as backup generators for sensitive loads (such as special manufacturing facilities, large information processing centers, hospitals, airports, military installations, large office towers, hotels, etc.)

for which long-duration energy supply failures would have catastrophic consequences. Engines and turbines currently account for most of the distributed generation capacity being installed—approximately 20 GW in the year 2000, or 10% of total capacity ordered. While nearly half of the capacity was ordered for standby use, the demand for units for continuous or peaking use has also been increasing.

The term *energy strategy* reflects the idea of accumulating energy in some form and then supplying it when needed. In its broadest sense, energy storage encompasses fuels such as coal, gas and uranium, as well as water reservoirs behind hydroelectric dams, or "cool" storage (chilled water in tanks) or hot water, all of which allow the controlled accumulation and release of energy. But the term *energy storage* is often used to refer specifically to the capability of storing electrical energy that has already been generated, and controllably releasing it for use at another time. This concept is sometimes called *electric energy storage* to distinguish is from other types of energy storage.

At present, electric energy storage in building-level distributed energy storage is most commonly associated with batteries that are used in conjunction with non-continuous power generators such as wind turbines and photovoltaic systems. These batteries are charged during periods of excess generation and then are discharged during periods of insufficient generation. Backup power systems such as small-scale UPS devices are also widely utilized. Moreover, there is increasing use of portable power storage devices; we find these in our computer, mobile phone, gaming systems, MP3 players, etc.

HOW IS DYNAMIC ENERGY MANAGEMENT DIFFERENT?

There is significant potential to increase the functionality of typical demand-side management measures, typical demand response strategies, and typical implementation of building-level distributed energy resources by combining them in a cohesive, networked package that fully utilizes smart energy-efficient end-use devices, advanced whole-building control systems, and an integrated communications architecture to dynamically manage energy at the end-use location. We refer to this concept here as *dynamic energy management*. Figure 7-2 illustrates the additional potential for functionality offered by dynamic energy management relative to conventional energy management practices. It transforms the energy-efficient end-use devices and processes associated with typical demand-side management into smart, highly energy-efficient end-use devices and processes. It transforms the standard distributed energy resources associated with typical practice into smart, environmentally friendly on-site energy resources that are leveraged to their maximum potential to benefit the end-user, the utility and the environment. It transforms the local, standard controls associated with conventional energy management into advanced, building-wide controls that are mutually compatible and capable of learning. Finally, it transforms the basic communications associated with typical demand response and typical distributed energy resources into an advanced integrated two-way communication architecture that network the end-use devices, distributed energy resources, and control systems with each other and with the utility or other external entities to dynamically manage and optimize energy use.

As discussed previously, a dynamic energy management system is a demand-side energy resource that integrates energy efficiency and load management from a *dynamic, whole-system* or *networked perspective* that simultaneously addresses permanent energy savings, permanent demand reductions. and temporary peak load reductions. This section briefly outlines the operation of a dynamic energy management system from an integrated systems perspective. Next, it specifies some of the characteristics that individual components of a dynamic energy management system are likely to embody. These components include (a) smart efficient end-use devices; (b) smart distributed energy resources; (c) advanced whole-building control systems; and (d) integrated communi-

Current Practice	Additional Potential
Basic Two-Way Communication	⟶
	Integrated Communications Architecture
Local Standard Control Systems	⟶
	Advanced Whole-Building Control Systems
Limited Distributed Energy Resources	⟶
	Smart Distributed Energy Resources
Energy Efficient End-Use Devices	⟶
	Smart Energy-Efficient End-Use Devices

Figure 7-2. Additional Potential of Dynamic Energy Management Beyond Current Practice in Energy Use Management

cations architecture. At the end, some of the key features of a dynamic energy management system that are likely to facilitate implementation and user adoption are discussed.

OVERVIEW OF A DYNAMIC ENERGY MANAGEMENT
SYSTEM OPERATION FROM AN INTEGRATED PERSPECTIVE

A dynamic energy management system is comprised of highly efficient end-use devices, equipped with advanced controls and communications capabilities that enable them to dynamically communicate with external signals and to adjust their performance in response to these signals. This marks an emergence from static to dynamic end-use devices with advancements in distributed intelligence. In addition to electric end-use devices, a dynamic energy management system would also include distributed energy resources such as solar photovoltaic systems, diesel generators and fuel cells. The performances of these distributed energy resources are also programmed to operate in an integrated manner with end-use devices at the facility so as to be able to optimize overall system performance. The energy-efficient end-use devices along with distributed energy resources are together referred to here as "smart devices."

These smart devices have a built-in programmable response and control strategy, whereby users are able to program the device performance and set optimal performance levels based on a variety of external parameters such as external ambient conditions, time of the year, consumer habits and preferences, etc. The devices are able to communicate with a variety of external signals such as electricity prices, emergency events, external weather forecasts, etc. For communication with external signals received from the energy service provider or with other external signals, advanced meters with communications infrastructure will be required. This will enable the energy service provider to connect the electric meter and end-use devices in the building to the Internet, thereby giving the energy service provider direct access and control to these devices.

Depending on the hourly electricity price or other external parameters and based on the pre-programmed control strategy, the smart devices equipped with the responsive controls automatically respond to the external signals and optimize entire system performance, say within the user "comfort range" to minimize electricity costs. For the dynamic energy management system to operate autonomously in response to electricity price or other external signals, it must be capable of very abstract decision making, ranging from determining the best cost vs. comfort tradeoff for current conditions, to the very physical, such as turning an air conditioner on or off. Information on different parameters such as temperatures throughout the building, outside weather conditions, occupancy, appliance use and power consumption may allow for targeted control and be able to deliver predictable behavior and energy cost to the occupants.

The response strategy of each individual smart device is networked and interacts with the response strategies of other devices in the system so as to be able to optimize entire "system" performance. Networking among devices allows internal communications and interactions among devices. The system should be able to execute a fully automated control strategy with override provisions. In a fully automated system, the control and communications technology listens to an external signal, and then initiates a pre-programmed control strategy without human intervention. The control devices have real-time control algorithms in their gateway devices and automatically control without manual operation. For example, they are enabled to respond to curtailment requests or high-energy prices from the energy service provider, and automatically deploy specific control strategies to optimize system operation and avoid high-energy costs.

Even though the system is able to control multiple devices, user preference may be to control some devices directly (e.g., HVAC), while for others (e.g., copiers and fax machines) direct control is probably not practical. For devices that are indirectly controlled, the actuation of response could be occupant-assisted through signaling the occupant via some kind of notification methods (e.g., red-yellow-green signals) that tell occupants when the time is propitious to run these appliances. Furthermore, an intelligent dynamic energy management system will contain learning functionalities with learning logic and artificial intelligence in order to be able to learn from prior experiences and incorporate these lessons into future response strategies. For example, if an occupant attempts to lower the temperature set point at a time when electricity is expensive, a confirmation will be required for the action to take effect. The fact that the occupant confirmed that a lower set point was desired even though it would be costly to achieve becomes part of the learning process. The measure of success for the learning functionality is that occupants override the system less over time.

KEY CHARACTERISTICS OF SMART ENERGY-EFFICIENT END-USE DEVICES AND DISTRIBUTED ENERGY RESOURCES (TOGETHER REFERRED TO AS "SMART DEVICES")

- Smart devices are comprised of very high-efficiency end-use devices and a variety of distributed energy resources (discussed in earlier sections).

- Smart devices are equipped with highly advanced controls and communications capabilities.

- Smart devices embedded with microprocessors will allow incorporation of diagnostic features within these devices based on critical operating variables and enable them to undertake corrective actions.

- Distributed energy resources with intelligent controls are able to synchronize their operation with end-use devices in order to optimize system performance. They are also enabled to automatically feed back power to the grid based on overall system conditions.

- Communications features for these devices need to be set up based on "open architecture" to enable interoperability.

- Smart devices contain microchips that have IP addresses that enable external control of these devices directly from the Internet or through a gateway. It is desirable to have TCP/IP[5] communication protocol so that the system can be set up and managed using common network management tools.

- These devices have "learning logic" built into them (artificial intelligence, neural networks) to improve on future performance based on past performance experience, and on parameters such as building cool-down and heat-up rates, occupant habits, outside temperature and seasonal variables.

KEY CHARACTERISTICS OF
ADVANCED WHOLE-BUILDING CONTROL SYSTEMS

Advanced whole-building control systems will need to incorporate the following functionalities:

- Receiving and processing information from sensors

- Sending actuation signals for control of devices

- Learning physical characteristics of the building from sensor information

- Managing time-of-day profiles

- Displaying system status to occupants

- Obtaining command signals and overrides from occupants

- Learning the preferences and patterns of the occupants

- Receiving and displaying external signals, such as price information from the utility

Advanced energy management and control system (preferably web-based) is likely to be an enabling technology.

KEY FEATURES OF A
DYNAMIC ENERGY MANAGEMENT SYSTEM

The key features of a dynamic energy management system can be summarized as follows:

Incorporate End-User Flexibility—Customers should have numerous options about how they can participate in a dynamic energy management system. Customer flexibility must be built into any state-of-the-art system. They should have the ability to use custom business logic that is applicable to their own operations. For example, in response to high electricity price signals, some may choose to allow remote real-time sheds, while others may want some advanced warning via pager or cell phone and the ability to opt out, if desired.

Simplicity of Operation Will Be a Key Features—For easy user adoption of a dynamic energy management system, the user interface for the system will have to be concise and intuitive for non-technical people. The system will need to work right out of the box with no programming requirement. It will need to behave autonomously based on effective initial defaults and machine learning.

Leverage on Standard IT Platforms vis-à-vis Building Custom Systems—One of the most important ways to keep costs low will be to leverage existing trends in technology. The use of existing IT technology in such systems wherever possible is an important way to keep costs low. The public Internet and private corporate LAN/WANs are ideal platforms for controls and communications due to their ubiquity, especially in large commercial buildings. In addition, the performance of IT equipment (e.g., routers, firewalls, etc.) continues to improve and equipment prices continue to drop. Dynamic energy management systems based on standard IT platforms will also tend to be more scalable and secure than special-purpose systems developed specifically for the purpose.

"Open Systems" Architecture and Universal Gateways Essential for Integrating System Operation—An important concept in the dynamic energy management system architecture is that the layers of protocols across all the systems are common, and that seamless communication and control activity can occur. An "open system" architecture is essential in integrating the system operation. Also, monitoring and controlling different devices from a central location or from anywhere in the network can be done only if a universal gateway is used. Communication between devices and the Internet is accomplished through standard communications pathways including Ethernet, telephone line or wireless communications.

Integration With Existing Building Energy Management Systems Essential—It will be advantageous for state-of-the-art dynamic energy management systems to have tight integration with any existing EMCS and EIS and enterprise networks within buildings. This strategy maximizes the performance, distribution and availability of the building data, while minimizing the installation and maintenance costs. In systems in which EMCSs and EISs are highly integrated with enterprise networks and the Internet, managing the flow of information is greatly simplified.

"Open Standards" and "Interoperability" are Key Characteristics—For flexibility and future proofing, state-of-the-art dynamic energy management systems should use open standards wherever possible. Unlike proprietary systems, truly open systems are interoperable. In other words, a device from one company (e.g., Cisco) will easily and naturally reside on a network with products from other companies (e.g., Nortel). Communication using the TCP/IP protocol will ensure that the system can be set up and managed using common network management tools.

Flat Architecture Essential for Robust, Low-Cost Systems—A state-of-the-art dynamic energy management system would have a flat architecture in which there are a minimum number of layers of control network protocols between the front-end HMI (Human Machine Interface) and final control and monitoring elements such as actuators and sensors. The most robust and least costly systems should have no more than one enterprise network protocol and one control network protocol.

CONCLUSION

In addition to the other features mentioned in this book, the smart grid must enable connectivity with customers in a dynamic systems concept.

References
Gellings, Clark W., *Demand-side Management: Volumes 1-5*, EPRI, Palo Alto, CA, 1984-1988.
Assessment of Demand Response and Advanced Metering—Staff Report, FERC Docket AD06-2-000, August 2006.
Dimensions of Demand Response: Capturing Customer Based Resources in New England's Power Systems and Markets—Report and Recommendations of the New England Demand Response Initiative, July 23, 2003.
Arens, E., et al. 2006. "Demand Response Enabling Technology Development, Phase I Report: June 2003-November 2005, Overview and Reports from the Four Project Groups," Report to CEC Public Interest Energy Research (PIER) Program, Center for the Built Environment, University of California, Berkeley, April 4.
Transmission Control Protocol/Internet Protocol (TCP/IP) as developed by the Advance Research Projects Agency (ARPA) and may be used over Ethernet networks and

the Internet. Use of this communications industry standard allows DDC network configurations consisting of off-the-shelf communication devices such as bridges, routers and hubs. Various DDC system manufacturers have incorporated access via the Internet through an IP address specific to the DDC system.

IntelliGrid Architecture Report: Volume I Interim User Guidelines and Recommendations, Report 1012160, Electric Power Research Institute, December 2005.

Fast Simulation and Modeling, Report 1016259, Electric Power Research Institute, December 2007.

Electricity Technology Roadmap: Meeting the Challenges of the 21st Century: 2003 Summary and Synthesis, Report 1010929, Electric Power Research Institute, June 2004.

Chapter 8

The EnergyPort^{SM} as Part of the Smart Grid

Electricity consumers are changing. Their behavior is increasingly driven by real-time management of electricity. In addition, they increasingly demand higher levels of reliability, efficiency and environmental performance. Electric systems increasingly are controlled by customer response and will be aided by real-time pricing and "dynamic systems."

Electricity is often taken for granted, but it is the building block which fuels the digital economy. The digital world is enabling a potential shift which could enable the electricity enterprise to shift from a commodity-based, risk-averse industry into an innovative, real-time creative industry.

Consumers do actually have unlimited control over their electricity use. However, they lack the will and the means to enable that control. The only information they receive is a month or more after they consume the energy. They do not have information on the time, pattern and amount of their usage. Nor do they have information on time-varying prices, options to purchase under alternative tariffs, or other electricity-related services.

Information and control are the key to enabling customers to manage their electricity costs. If customers and their appliances have the ability to make consumption decisions, price-driven incentives will encourage customers to use electricity more judiciously—using less and managing peak demand. While a segment of today's customers may be willing to undergo the inconvenience of monitoring the electricity market in real-time and adjusting the operation of appliances and devices to minimize costs—that segment is in a minority. Many more customers would be interested in an automated system which would automatically monitor prices and system conditions and adjust individual building, process and appliance systems to respond.

If the systems are in place, then signaling customers appropriately—particularly that peak prices are higher than off-peak prices—will result in a change in consumer behavior. Demonstration programs confirm that customers do not want to spend a great deal of time managing their energy use, but some segments want to know more about electricity consumption patterns.

There is evidence to suggest that the combination of ubiquitous low-cost communications, wireless and wired internet protocol (IP) addressable, standardization, low-cost sensors make precise, real-time and on-demand electricity management a low-cost increment to investments already being made to serve other needs (e.g., security, web connectivity). As a result, many new firms offer new customer-centric products and services. This may create a "market pull" which could fuel the need for regulators to adopt new regulatory arrangements which will allow utilities to invest in advanced metering infrastructure and energy efficiency.

These dynamic systems will allow the entire electricity infrastructure to respond to demand changes, allow for much more efficient decisions on where electricity is generated, and how it is distributed and utilized. This would enable the electricity enterprise to operate more economically, will improve the environment, reduce the needs for energy overall, and encourage investment.

While some of the incentives to engage with customers exist now, the principal incentives lie with a potential change in the regulatory compact. In most jurisdictions, investor-owned electric utilities are able to deliver effective and economic electricity energy. They are not, however, able to offer comparable compensation to stakeholders for investments in most demand-response programs and in any energy efficiency programs.

However, in many cases, load-serving entities do have an incentive to reduce peak demand since the production and purchase of electricity at those key times, especially during peak periods, lowers costs. In addition, utilities have limited incentive to minimize distribution throughput.

Most utilities have the opportunity to replace today's mechanical electric meters with a smart metering infrastructure that is capable of providing the customer with information using advanced meters to reduce costs and enhance services. Advanced metering can be adapted to provide customers with more information and allow the customer to start making decisions.

Customers routinely make investments in home and commercial electronics for reasons other than managing electricity—heating, ventilation and air conditioning (HVAC) control, home entertainment, security, and other services. As sensors and information technology networks become more common, the cost of retrofitting existing buildings will decline and new buildings will be designed to accommodate control and implementation of dynamic systems.

Over the next decade, the adaptation of advanced electricity management will become an integrated part of the entire electricity value chain.

As the world drives to become increasingly digital, and communications computational ability and customer-enabled energy management control becomes ubiquitous, the electricity industry is an anomaly. It is a centralized supply-driven industry built around a philosophy of cost recovery. In the evolution of the electric utility industry, its focus is on the regulatory compact and the utilization of assets in generation, transmission and distribution to serve customers interior need for electricity. Unfortunately, the compact created then did not include the entire electricity value chain. Electricity is produced by one or more forms of energy and then delivered to customers. The traditional view of that system ignores the ultimate conversion of electricity into useful energy service. Therefore, the utility industry remains a highly asset incentivized industry. In recent years, an independent merchant generation business dominated the supply business.

The system of regulation has served us well and established the basis for building the electricity enterprise. It has performed extraordinarily well. By its nature, it is very command and control. It relies on top-down planning, decisions and operational control. It is not designed to enable "utilities" to invest on "both sides of the meter" to ensure the entire production-delivery-utilization chain is optimized. The "customer side of the meter" is perceived as someone else's problem.

The assumption is made that consumers will make the right decision regarding prices, the availability of end-use energy consuming devices and their control of those devices. In fact, research has shown that consumers do not choose the most optimal or most efficient appliances or devices, they do not necessarily participate in demand-side management programs, and their control of appliances and devices is varied.

It is interesting to note that in many industries there has been a transformation to distributed systems. In electricity, the shift to distributed resources could be an evolutionary event.

The modestly successful effort of the 80s and the 90s to create a demand-side management (DSM) structure has impacted the industry's shift to a customer-centric approach. Unfortunately, the reduction in consumer acceptance and response of demand-response programs is cited as evidence. The Department of Energy estimates a one-third decline. With higher energy prices, there is a greater willingness to look at demand-response and energy efficiency efforts. Yet this willingness may still not provide sufficient sustainable incentives to make the necessary changes in consumer behavior.

Utilities remain cautious about energy efficiency and demand response, even with regulatory incentives. They cite uncertain DSM results; get involved in hardware and services to customers; lack of communications infrastructure and of common protocols; and a lack of experience in joint venturing and teaming with retailers.

Early efforts at energy efficiency received bad press because of poor design, imperfect pricing, awkward administration, and unsatisfactory monitoring, even though those programs have been an overall success.

Unfortunately, utilities and other electric service providers have had bad experiences investing in customer-related technologies such as telecommunications, internet technologies, home and building security services, and other customer-related services. To the extent that some utilities were investing in a communications network, they were often not doing so using open IP protocols.

These uncertainties have created a resistance to any new investment in the customer side of the value chain. Surveys have revealed that there is near universal agreement that working with electric utilities is difficult, both due to their reluctance to engage with customers and a resistance to investment in any technologies that reduce kilowatt hour sales.

In most states in the U.S., the regulatory compact given to electric utilities encourages them to build adequate generation, transmission and distribution capacity to meet their consumers' demand for electricity. They receive a return on that investment subject to limits also adjudicated by the state commissions. Signs exist that this so-called "compact" may be renegotiated. This would unleash innovation in electricity retail space.

The result is that today, few consumers of electricity are incented to or able to manage their electricity demand—they have no price signals, no choice among usage profiles that reflect their social and economic views, and no portals or interfaces that allow them to see how they are using electricity. However, this status may evolve. With the evolution of web-enabled energy management and cheap, ubiquitous communications (wired and wireless), there is the potential to piggyback on the "computer revolution" to build on infrastructure and cultural affinities incrementally. The question is whether today's incumbent electric utilities will leverage this opportunity or leave it to third parties.

There is some evidence to indicate that a segment of electricity consumers will want to exercise control of their electricity use—if the technologies were present which facilitated that control. Many will not be engaged in the heavenly pursuit of managing the operation of their energy-consuming devices and appliance and the building they are resident in. The segment of the population willing to engage in energy management will grow as more are convinced as to how this is in their enlightened self interest.

Engaging in control over one's electricity destiny may not mean sitting in front of a PC all day. It may mean setting a few controls on a "master controller" so as to have a brain automate the purchase decision for the consumer.

The electricity industry is moving toward an intersection where higher electricity and fuel prices and changes in electricity rates are aligning with dramatic advances in real-time internet protocol (IP)-based wireless communication. These include IT advances, especially IP-based and inexpensive sensors; heavier reliance on high-quality electricity in home-entertainments backup and standby devices; emerging mass market in residential wireless sensor management; and marginal pricing which sets prices high enough to stimulate innovation and the penetration of new technologies.

It is possible to imagine a future where virtually all electricity-using devices incorporate sensors that manage the pattern and amount of consumption of the consumer. This may include the capability to interface with the energy management systems resident in industrial facilities and large commercial buildings. Through electricity portals, customers will be able to program energy management capabilities. In addition, these advanced systems will allow interactive control in response to signals for utilities and other providers.

As a result, the power system will be able to respond to demand changes; it will allow for more efficient decisions regarding where electricity should be generated and how it is distributed. The result will be a more economical, environmentally friendly system which will encourage investments in perfect power systems.

Tomorrow's power system will incorporate electric energy storage capacity and a micro-grid orientation that will allow consumers demand to be managed at the substation- and-below-level to maximize the power system. In the future, it is possible that a significant portion of new generation could be built locally utilizing various distributed and renewable resources. There will be significant variations in deployment.

Some researchers believe that within 10 years the penetration of new technology will be underway as an increment to the current trends in home and business automation which will allow rapid deployment of electricity-related management technology. It is essential that structural incentives are put in place to supersede today's utility opposition and reticence to consumer programs. These technologies will create the potential for radically changing the electricity system from a supply-side system to a high-reliability "perfect" demand-driven incentive industry.

There is evidence to suggest that a segment of electricity consumers will take advantage of the ability to manage their electricity once that ability is enabled. This includes enabling consumers to be aware of the variability in electricity demand and price. The demand for electricity varies daily and seasonally. The daily variation ranges from lower demand overnight, a rise in demand in the morning to a moderate period through the day, a high-demand period in the late afternoon and early evening, and a return to a lower, moderate demand in the evening. Consumers cause this pattern of demand due to their use of energy-consuming devices and appliances. In part, these patterns are based on household schedules. To some extent, they are also caused by daylight and weather, i.e., less artificial illumination is required on bright days, more air conditioning on hot days, etc.

Meanwhile over these periods of usage, electricity costs vary—sometimes appreciably. Yet the consumer has no knowledge of those variations. And furthermore, no incentive to modify either the pattern or amount of demand. In short, the prices individual consumers pay bear little or no relation to the actual cost of providing power in any given hour. The price of electricity in most markets is based on a combination of long-term and short-term markets. The long-term purchases are negotiated and set by contracts. The short-term purchases are made on the basis of a variety of market-based exchanges. Typically, wholesale electricity is sold at a combination of fixed and marginal costs, but the customer does not feel the affects of those prices until they receive their bill some weeks later. As a result, they see no connection between their behavior and the cost of electricity. Consumers have no incentive to change their consumption as the cost of producing electricity changes. The consequences of this disconnect between cost and consumption yields inappropriate investments in generation and transmission.

This arrangement could be radically modified by the implementation of a dynamic system—a system which would communicate the price of electricity and enable consumers to respond to those price signals. Dynamic systems harness the recent enhancements in information technology. These same technological developments also give consumers a tool for managing their energy use in automated ways. Customers can make their own pricing decisions, use default profiles, or let the supplier control demand through switching.

The evidence of the past 20 years suggests that electricity consumers will respond to time varying prices. However, the focus is shifting to the question of the symbiosis of pricing and technology. In conjunction with time-varying pricing signals enabled by a dynamic system, the ability of customers to choose and to control their electricity consumption using digital technology is at the core of transforming the electric power system.

Dynamic systems and the digital technology that enables them are synergistic. Dynamic systems can enable a diverse and consumer-focused set of value-added services. Dynamic systems empower consumers and enable them to control their electricity choices with more granularity and precision. This can range from energy management systems in buildings that enable consumers to see the amount of power used by each function performed in the building to appliances that can be automated to change behavior based on changes in the retail price of electricity.

Dynamic systems can yield lower wholesale electricity prices, better asset utilization and reduced needs for additional generation and transmission investment.

Increased reliability is one particularly valuable benefit of implementing dynamic systems. Active demand response to price signals inherently acts to moderate strains on the entire system. Dynamic systems allow the reduction of peak-period consumption reducing the likelihood of transmission congestion and generation shortages.

One of the most important benefits of dynamic systems are their ability to unleash innovation. The ability for consumers to be informed about their electricity behavior creates incentives to seek out novel products and services that better enable them to manage their own energy choices and make decisions that better meet their needs. This incentive, in turn, induces entrepreneurs to provide products and services the consumers demand. Competition for the business of customers would drive innovation in end-use technologies.

The EnergyPort would facilitate consumer engagement with such dynamic systems.

WHAT IS THE ENERGYPORT[SM]?

The EnergyPort[SM] is part "portal," part "gateway," and part "smart electric meter." To date, the term portal has been narrowly defined as relating to communications data hubs, gateway often refers to specific, narrow access points, and meter is the traditional electric utility interface with its customers limited to measuring consumption. In fact, the term "portal" is experiencing growing use in reference to internet-based web servers that provide a web-based view into enterprise-wide activities. The EnergyPort[SM] provides a view into consumer facilities and carries the definition further to include communications with energy management systems and even end-use subsystems and equipment. The EnergyPort[SM] is a device or set of devices that enable intelligent equipment within consumer facilities to communicate seamlessly with remote systems over wide-area networks. The EnergyPort[SM] can perform the functions of a communications gateway that physically and logically inter-connects a wide-area access network with a consumer's local network. It is envisioned that the EnergyPort[SM] will have locally available computing resources to support local monitoring, data processing, management, and storage.

WHAT ARE THE GENERIC FEATURES OF THE ENERGYPORT[SM]?

The EnergyPort[SM] may benefit consumers in seven ways: safety, security, convenience, comfort, communications, energy management, and entertainment.

Simplify Building Systems

Today's buildings use separate systems for ethernet, power, telephone, intercom, thermostat, cable TV, audio and video distribution, security wiring, doorbells, and the like ... some wireless and others often in separate wiring systems.

In the EnergyPort^{SM}, those systems, and others, will be simplified. Three approaches will be available: (1) They can be integrated and consolidated into one hybrid cable. That cable will run to every convenience outlet in the building and provide all the power, communications, and control capabilities the building owner will need at any convenience outlet in the building; or (2) power line carrier signals will be superimposed over the buildings power lines; or (3) radio networks will send signals displacing all but the basic electrical wiring.

No longer will consumers have to search for special jacks for audio, video, or telephone. No longer will they have to run separate lines for security systems or intercoms. The EnergyPort^{SM} System can plug printers, fax machines, DVDs, VCRs, toasters, telephones, radios, TVs, stereos or speakers into a universal outlet that will link them to each other and to the services they require.

The use of such a universal outlet will not preclude the use of any conventional appliance. Any of the old, conventional appliances consumers may wish to use in the EnergyPort^{SM} System will be immediately usable with the system. The EnergyPort^{SM} System is an *enabling* system that handles the *distribution and control* of electric energy and other energy sources as well as all kinds of communications.

Safety

The EnergyPort^{SM} System can enable the distribution of electric energy far more safely through the use of a "closed-loop" system. Before energy can be fed to any type of outlet, an appliance must be plugged in and it must actually request power through a small communication chip it will contain. A microprocessor will identify each appliance to the EnergyPort^{SM}. Unless that happens, no energy is available at the outlet. Because the outlet is effectively "dead," the probability of electrical shock or fire hazards are substantially reduced.

In such a "closed-loop" system when an appliance is plugged in, its communications chip supplies an appliance identification to the system (much like the bar code does at the grocery store). It also provides an estimate of the appliance's energy consumption, and lets the system know the appliance's status—is it on or off, is it in proper working order. Only then is energy allowed to flow through the outlet to the appliance. And then its use is continually monitored by the system.

Now if that monitoring signal is interrupted or terminated, energy flow is immediately shutdown. Reasons for interruption might include unplugging the appliance or shutting it off; it could be an overload, short, or ground-fault condition; or it might be a poorly made plug connection or a short. Regardless, the consumer remains safe and secure because the energy flow has been shutdown … and will not be able to start again until the proper conditions exist.

Reliability

The EnergyPort^{SM} could offer even more to consumers in the event of an outage. There could be several parts of the system that use powered control devices. In order to keep these devices alive when the main power fails, an uninterruptible power supply can be enabled based on a storage device. This storage system could allow the system to keep intact any instructions programmed in; it could continue any timing operations necessary; and it could provide the necessary minimum power for security and alarm systems.

Another advantage of the EnergyPort^{SM} system that can come into play when there's a power outage is the integration with alternate power sources, such as photovoltaics, a standby generator, or gas cogeneration systems, the EnergyPort^{SM} will immediately switch over to the appropriate source.

Depending on the sizing of the alternate power source, the EnergyPort^{SM} System could provide the power necessary for selection of operations, such as the freezer, or refrigerator, cooking or lighting. Not only is this capability convenient, but in some home situations, it could be critical.

Decentralized Operation

The EnergyPort^{SM} System is not a computer-controlled building in the sense that there's a single personal computer running things. The EnergyPort^{SM} uses distributed intelligence, or distributed control, to carry out its operations. The reasons for this are efficiency and reliability. Much like good management in a business organization, decisions and actions are carried out most efficiently at the lowest levels possible. Thus, consumers don't have the president making every decision or doing every task.

In the EnergyPort[SM] System, the reliability of the entire building energy and communication system is integrated. So when there's a failure of any one component, most problems can be isolated and kept from affecting other operations because of distributed control.

Consumer Interface

Consumers will interact with the EnergyPort[SM] in a variety of ways. Everything in tomorrow's buildings will be controlled by signals... whatever device can send the appropriate signal can be used to control whatever the consumer chooses. No longer is the switch on the wall physically opening and closing the path of power to an outlet or fixture. Now it is simply sending a signal to the system. Therefore, that switch can be assigned to control anyone of a range of desirable applications. The same is true for a multitude of other devices, such as display panels, video touch screens, telephone keypads, infrared remote control devices, or even vice recognition devices.

For example, from switch or panel display, consumers may be able to select to control the lights, music, temperature, time, or the TV. To control the TV, perhaps these options now become available: beyond just simple on/off, access to control the volume, color, contrast, and channel.

In homes today, there is a proliferation of appliances, each with its own set of instructions. For example, to set the timing of an operation, you might use one procedure for the washing machine, another for the DVD or VCR, another for the microwave, and still another for the alarm clock. However, one of the neat things about EnergyPort[SM] control is that whatever the consumer uses to control operations or timing, they only have to learn one basic set of instructions. Keeping it simple is a basic premise of the EnergyPort[SM].

For example, the living room might be set up to have the table lamp controlled by either the telephone or a wall switch. The floor lamp might be assigned to an occupancy detector as well as another wall switch. If the furniture was rearranged or it was more convenient to run operations differently, the house would not have to be rewired, the assignment would simply be changed.

Sensors are just another kind of switch. And their use can be very helpful in the EnergyPort[SM] System. In this case, they can be used to determine unauthorized entry, and occupants can decide what response should be made. It might be to light up the outside at night, turn on all the interior lights, sound an alarm, and call the police of an emergency monitoring service for assistance.

Those same sensors can do double duty. During the day, when consumers are at home, they might simply determine an individual's presence so that they can light their way from room to room. Consumers could carry the laundry, or the baby, from room to room without having to fumble and stumble for a light switch. In the EnergyPort^SM, sensors can also work interactively. For example, why should the lights be turned on in a room that already has enough light coming in the window on a bright day? Or why should the lights come on if no one's in the room at all? The EnergyPort^SM can enable remote control whether by a switch or telephone keypad or the like. This can come in very hand in order to issue one command from the comfort of one's bed at night to lock the doors, turn off all the lights, set the security alarms, and turn the thermostat to its set-back position. Telephone control can come in handy when unexpected events cause a change in plans. Here, it could prevent a char-broiled steak for dinner from becoming charcoal. Voice recognition devices are becoming increasingly reliable and affordable. Such a unit in the EnergyPort^SM System can make life better for all of us, but particularly for the elderly or infirmed who might not be able to exercise other forms of control.

Appliances That Talk to Each Other

One of the most exciting forms of control made possible by the EnergyPort^SM comes through appliance-to-appliance communication. Here, for example, the ringing of the telephone or doorbell would cause the vacuum cleaner to shutdown so a consumer could hear and respond.

Safety

Safety has a high priority among consumers. The EnergyPort^SM would monitor safety. Another form of safety comes when we can prevent dangerous things from happening ... as in the case of a young person who arrives home from school before his/her parents have returned. To keep the student from playing Betty Crocker in an unsupervised situation, simply lock out the use of the range until parents return. The same can be done for power tools in the workshop, an expensive stereo system, or a sophisticated sewing machine. Still another form of safety comes when consumers can be kept from making mistakes ... such as leaving the range burner on.

Fire, of course, is something to always be mindful of. Should one break out in a building enabled by the IntelliGrid^SM, the occupants can be alerted, told where the location of the fire is, what rooms are occupied, and flash the lights showing the safest pathway to exit the home. Meanwhile, all this information can be called out immediately to the fire department or emergency monitoring service so they can respond promptly and appropriately to the problem.

The use of sensors in a building need not involve cumbersome extra wiring or complications. Indeed, the convenience of being able to plug in whatever kind of sensor can meet the consumer's needs into an outlet conveniently located anywhere significantly reduces costs and adds a great deal to the safety and security of the home or business.

In addition, the EnergyPort^SM provides the convenience of being able to plug in any appliance anywhere in the home. No longer will consumers have to run unsightly, and sometimes unsafe, extensions for power, cable TV, telephone, or stereo when they choose to rearrange the furniture. Right now, for example, if a particular light switch is wired to a particular outlet where there is a table lamp, an extension cord has to be run to control it if the consumer decided to place it elsewhere. With the EnergyPort^SM System, the system can be reconfigured to control a lamp regardless of location.

Another kind of convenience comes when consumers are visited with a permanent or temporary disability or simply a change in the way they will be using their home or building. Should a particular product or control device be needed for a period of time, they simply go out and purchase or rent the EnergyPort^SM component to meet their needs.

Comfort is a key factor in deploying the EnergyPort^SM System. The EnergyPort^SM will enable consumers to set different temperature and humidity levels in different areas of their buildings. For example, keep the temperature cooler in a room where Thanksgiving dinner is being prepared, while it remains comfortably warmer for guests in the living or dining room. And why heat the bedrooms during that time just to keep guests' coats warm? The EnergyPort^SM System will allow separate zones or rooms of the home to be conditioned to suit specific needs at any time of the night or day.

Communication

Essential to the EnergyPort^SM are the deployment of advanced communications capabilities. Whether it be to coordinate all the items on the menu for a particular meal so that they all are finished at the same time for dinner ... or simply to use the telephone as an intercom to keep you from running or shouting to one another throughout the building. Perhaps the capability of placing a small video camera in the nursery will give a consumer more peace of mind when the babysitter is down in the family room munching popcorn and watching TV ... or being alerted to a problem in one of the appliances in a building. A problem, that if undetected, might have a cost the consumer more money or caused a bigger problem.

Not only can the EnergyPort[SM] System relay diagnostic problems from appliances, it can also let building owners know if the system itself is suffering any problems. Or, perhaps, just keep building owners apprised of its status or maintenance requirements, so you can respond appropriately.

The EnergyPort[SM] can help consumers manage the use of energy more effectively. It will help building owners and operators decide when to use heating and cooling systems based on whether or not anyone's occupying the space, what rooms are occupied, or what the cost of energy is at a particular time.

Heating water is one of the more costly uses of energy. Why spend the money to maintain a constant temperature when a particular level of hot water is only required at certain times. Also, there can be different water temperatures for different needs. No longer does heat have to be used for the worst-case situation. Now consumers can have their shower water, dish water, or clothes water at the right temperature at the right time.

In "closed-loop" control, the communications chip in each appliance was described as also providing energy consumption information for that appliance to the system. The feature enables the building owner to call up a printout or display of his estimated energy use whenever wanted. That estimate can be broken down appliance by appliance, time of day, duration of use, and cost of use. This is a much sought-after feature that places the consumer in control of how energy expenses are managed.

Of course, many consumers are not inclined to use such information. They'll simply want to make some preliminary decisions about how they'll spend that energy dollar for different degrees of efficiency, comfort and convenience. Many will want to take advantage of "real-time" pricing structures likely to be introduced by more and more utilities around the country. By so doing, the consumer can save money by time-shifting their use of energy. It may be as simple as cycling the air-conditioner to save energy during the cooling season. For the utility companies, the EnergyPort^SM System enables demand-side load management; for the consumer, it can mean significant savings without sacrifice.

Another help to both the utilities and the consumer comes from what the EnergyPort^SM can do when power must be restored after an outage. Because each building can be linked to the utility company, the utility can now selectively restore the power to appliances in the home. This means the utility doesn't have to wait until it has enough power available to handle the huge start-up current demand that's called for when power is restored all at once. Therefore, the utility can restore full power sooner to affected consumers ... and the consumer doesn't have to be without power for as long as he might have been.

Entertainment

Entertainment is an additional benefit of the EnergyPort^SM. The EnergyPort^SM can enable one central audio source to provide a selection of music to several different rooms of the home. Someone can listen to punk rock in the rec room, while teenagers can be soothed by Beethoven in the bedroom ... or vice versa!

Network Communications Management

The EnergyPort^SM can support both the applications that are dependent on internetworking as well as the functions necessary to manage the networks and connected devices.

Remote Consumer-Site Vicinity Monitoring

The EnergyPort^SM can support local data monitoring to support a variety of applications including, but not limited to, electric, gas and water operations support, quality-of-service monitoring, outage detection, and other functions that can support the provision of reliable digital quality power for the future. Such monitoring capabilities could be extended to other parameters such as security, microclimate weather, and home energy.

Markets

One of the greatest values of the EnergyPort^SM can be to enable consumers to effectively respond to electric energy market dynamics and real-time pricing. Today's metering doesn't enable consumers to respond to prices that vary hourly or even more frequently. Communicating market pricing to consumer-owned energy management systems and intelligent end-use equipment will greatly help to close the gap between consumers and market pricing for electricity.

CONCLUSION

The smart grid would be the glue which enables the ElectriNet[SM] to stick together. It would connect Low-Carbon Central Station Generation to local energy networks and to electric transportation. Within the local energy network, some form of portal would be needed to enable the ultimate connectivity—that to the consumer.

References
EnergyPort[SM] Features—Opening a Gateway to the World, C.W. Gellings, A Bold Vision for T&D, Carnegie Mellon University, December 2004.

Chapter 9

Policies & Programs to Encourage End-use Energy Efficiency*

In order to maximize energy efficiency improvements, high level leadership in the development, implementation, and monitoring of policies and programs is required by international government bodies and organizations, national governments, state governments, regional coalitions, energy companies, city governments, corporations, etc. Appropriate strategic plans need to be devised, and realistic but aggressive targets need to be established. The effectiveness of specific policies and programs will be a function of their design, stringency, and level of implementation. Ultimately, gaining the cooperation and dedication of *individual* energy end-users will be essential to produce the magnitude of response necessary to achieve targets for energy savings.

There are many proven instruments for achieving energy efficiency results. Some are regulation-driven and others are market-driven. Table 9-1 lists a variety of policy- and program-instruments that have been used successfully and have the potential to yield significant energy efficiency improvements. The table organizes the energy efficiency instruments into five categories:

- General
- Energy supply & delivery
- Industry (including agriculture and waste management)
- Buildings
- Transport

The general category includes policies and programs that broadly apply to the economy as a whole. Specific instruments include subsidies

---

*Based in part on material prepared by Clark W. Gellings and Patricia Hurtado, Kelly E. Parmenter and Cecilia Arzbaeher of Global Energy Partners, LLC.

Table 9-1. Examples of Policies and Programs for Achieving Energy Efficiency Improvements

General	Energy Supply & Delivery	Industry	Buildings	Transport
Subsidies for research and development	Minimum standards of efficiency for fossil-fuel-fired power generation	Strategic energy management	Stronger building codes and appliance standards	Improved fuel economy
Public goods charge to fund energy efficiency programs	Reduced fossil fuel subsidies	Energy efficiency standards, including standards for advanced motors, boilers, pumps, compressors, etc.	Labeling and certification programs	Mandatory fuel efficiency standards
Tax incentives	Carbon taxes or charges for fossil fuels		Advanced lighting initiatives	Advanced vehicle design and new technologies
Incentives for private sector investment	Innovative rate structures	Process-specific energy efficient technologies	Leadership in procurement of energy efficient buildings	Taxes on vehicles, fuel, parking, etc.
Leadership in procurement of energy-efficient equipment, vehicles, and facilities	Decoupling of profits from sales to encourage energy efficiency	Energy management systems	Energy efficiency obligations and quotas	Mass transit improvements
	Mandatory demand-side management programs (including energy efficiency and demand response measures)		Energy performance contracting	Transportation infrastructure planning
Public information and education to increase awareness	Meeting larger portion of future demand with energy efficiency	Negotiated improvement targets	Voluntary and negotiated agreements	Telecommuting
	Greater use of combined heat and power	Incentives	Tax incentives	Mode switching
	Improved infrastructure (including smart grid)	Research initiatives	Subsidies, grants, loans	Leadership in procurement of energy efficient vehicles
Trade allies	Reduced natural gas flaring	Benchmarking	Carbon tax	
	Tradable certificates for energy savings		Education and information	
			Mandatory audit and energy management requirement	
			Detailed billing	

Sources:

1. *Realizing the Potential of Energy Efficiency, Targets, Policies, and Measures for G8 Countries,* United Nations Foundation, Washington, DC: 2007.

2. *Climate Change 2007: Mitigation,* Contribution of Working Group III to the Fourth Assessment Report of the Intergovernmental Panel on Climate Change, B. Metz, O. R. Davidson, P. R. Bosch, R. Dave, L. A. Meyer (eds), Cambridge University Press, Cambridge, United Kingdom and New York, NY, USA: 2007.

3. *Energy Use in the New Millennium: Trends in IEA Countries,* International Energy Agency, Paris, France: 2007.

for research and development; public goods charges to fund energy efficiency programs; tax incentives; incentives for private sector investment; leadership in procurement of energy-efficient equipment, vehicles, and facilities (e.g., by government or high-profile corporations); public information and education to increase awareness (such as via mailings, media, direct contact, etc.); and trade allies.

The energy supply & delivery category encompasses actions directed at energy companies. Some of the potential actions are development of minimum standards of efficiency for fossil-fuel-fired power generation; reduced subsidies for fossil fuel; carbon charges; innovative rate structures including those that decouple profits from sales to encourage energy efficiency; implementation of demand-side management programs; targets of meeting a larger portion of future electricity demand with energy efficiency; greater use of combined heat and power; improved power supply and delivery infrastructure (including development of a "smart grid"); reduced natural gas flaring to prevent waste of valuable resources and to reduce emissions; and introduction of tradable certificates for energy savings. The concept of tradable certificates for energy efficiency is analogous to approaches taken for green certificates and CO_2 trading, but applies to energy savings and meeting energy efficiency targets.

The industry category consists of energy efficiency instruments such as strategic energy management; minimum standards of energy efficiency for key types of equipment including motors, boilers, pumps, and compressors; process-specific energy efficient technologies; energy management systems; negotiated improvement targets; incentives; research initiatives; and benchmarking of energy efficiency. Some of the policies and programs extend to the agriculture and waste management categories.

There are many policy and program opportunities in the buildings category. Some of the main opportunities include stronger building codes and appliance standards; labeling and certification programs; advanced lighting initiatives aimed at phasing out inefficient lighting and promoting energy efficient lighting and advanced controls; leadership in procurement of energy efficient buildings; energy efficiency obligations and quotas; energy performance contracting, in which an energy service company pays the capitals costs of the energy efficiency measures and is paid back by the energy savings; voluntary and negotiated agreements; tax incentives; subsidies, grants, and loans; carbon tax; education and information; mandatory audits and energy management requirements; and detailed billing.

Policy and program instruments applicable to the transport category relate to vehicle and fuel efficiency as well as to shifting transportation modes, routes, and schedules. Specific instruments are improved fuel economy; mandatory fuel efficiency standards; advanced vehicle design and new technologies (e.g., plug-in hybrid electric vehicles); taxes on vehicle purchases, fuel, and parking; mass transit improvements to make it more desirable to the general public; infrastructure planning to optimize routes, traffic flows and general efficiency; greater allowances for telecommuting; mode switching to more efficient forms of transport; and leadership in procurement of energy-efficient vehicles.

In addition to the policies and programs listed in the table, there are also potential policies related to aiding transition and developing nations. For example, the United Nations Foundation report suggests establishing loan guarantee funds for investments in efficiency; making investments in the human and institutional resources required by the countries to maximize efficiency; supporting a market exporting energy efficient technologies; and minimizing the trade of inefficient or lesser-efficient technologies (UN Foundation, 2007). The most successful policies and programs, particularly for developing countries, are likely to be those that address energy requirements for economic development while simultaneously considering environmental impacts and costs.

POLICIES AND PROGRAMS IN ACTION

The following sub-sections present nine examples of the types of energy efficiency policies and programs currently underway at the multi-national, national, state, city, and corporate levels. Example 9-1 is a multi-national-level example, Examples 9-2 to 9-6 are national-level examples, Example 9-7 is a state-level example, Example 9-8 is a city-level example, and Example 9-9 is a corporate-level example.

Multi-National Level
In 1999, the International Energy Agency (IEA) proposed that all countries harmonize energy policies to reduce standby power to 1 Watt or less per device in all products by 2010. Further, the IEA proposed that all countries adopt the same definition and test procedure, but that each

country use measures and policies appropriate to its own circumstances.

While the standby power use for most small devices is relatively small, typically ranging from 0.5 to 10 Watts per device, the total power use of all devices drawing standby power is substantial because of the current and even higher expected proliferation of devices. Indeed, it is estimated that worldwide standby power currently accounts for 480 TWh each year. IEA estimates standby power use could be reduced by as much as 60 to 80%.

To date, there have been several accomplishments related to the 1-Watt Initiative:

The G8 has committed to promoting the 1-Watt Initiative.

An internationally sanctioned test procedure for standby power was adopted by the International Electrotechnical Commission (IEC 62301) in 2005. Today, this procedure is specified and used extensively.

Many countries around the world use voluntary endorsement labels.

Korea and the U.S. have implemented government procurement.

Japan and California are currently the only two regions that have adopted regulations. However, standby requirements increasingly are part of wider energy efficiency regulations. For example, Korea, Australia, New Zealand, the U.S., Canada, and China are considering regulations.

The voluntary Code of Conduct in Europe has been expanded to cover standby power in external power supplies, set-top boxes, and broadband modems.

The primary lesson learned from the 1-Watt Initiative is that it is difficult and expensive to target policies towards individual devices. As a result, the IEA now proposes a uniform approach to standby power requirements in all products. This would apply to all products with the exception of products already regulated by an efficiency standard with a test procedure that captures standby power use, or products with special features such as medical devices.

Sources:
1.   Ellis, M., International Energy Agency, *Standby Power and the IEA*, Presentation, Berlin, May, 2007.
2.   *Standby Power Use and the IEA "1-Watt Plan,"* Fact Sheet, International Energy Agency, Paris, France: April 2007.

National Level: Norway

In 2002, the Norwegian Ministry of Petroleum and Energy (MPE) established an energy efficiency agency: Enova. Enova is funded from a levy on the electricity distribution tariffs. Its 2007 budget is $200 million. Enova's mission is to promote energy efficiency and renewable energy generation in a consistent and comprehensive manner in Norway. The goal of Enova is to achieve reductions in energy use of 12 TWh by 2010, either through improved end-use energy efficiency or increased production of renewable energy. (Total net energy use in Norway was 222 TWh in 2006. Thus, Enova's savings goal of 12 TWh represents ~5% of total energy use in Norway).

Enova primarily relies on improvements in industrial energy efficiency to achieve its goal. Specifically, Enova provides investment aid, or grants, to industry. These grants may cover as much as 40% of the investments costs. (The maximum of 40% is imposed by state guidelines.) Enova uses a standard net present value approach to assess the profitability of a project. The main criterion used by Enova for selecting energy efficiency projects is the project's investment aid per unit of energy saved. The lower the investment aid per unit of energy saved, the more likely the aid will be granted. However, the grants have some obligations attached. For example, if the energy savings are not met, parts or all of the investment aid may have to be returned. Similarly, if the actual investment costs are lower than the estimated investment costs, the aid is reduced proportionally. On the other hand, if the actual investment costs are higher than the estimated investment costs or the energy savings exceed the estimated energy savings, the amount of the grant still remains the same. As a result, the inherent incentives for a company to signal high investment costs and high energy savings are somewhat reduced.

The results from the Norwegian energy efficiency investment model are promising. By the end of 2006, Enova had contractual agreements for an aggregated savings of 8.3 TWh/year. Contractual agreements associated with industrial energy efficiency projects accounted for 2.2 TWh/year, or 26% of total aggregated energy savings. Numerous companies use the results from the energy efficiency projects to promote their commitment to corporate social responsibility.

Sources:

1.   Enge, K., S. Holmen, and M. Sandbakk, *Investment Aid and Contract Bound Energy Savings: Experiences from Norway*, 2007 Summer Study on Energy Efficiency in Industry, White Plains, New York: July 24-27, 2007.
2.   Statistisk Centralbyra, *Statistics Norway, Total supply and use of energy, 1997-2006*.
3.   *Enova's results and activities in 2006*, Enova: 2007. (In Norwegian).

National Level: Ghana

African countries primarily focus on energy-efficient and sustainable supply-side energy initiatives. For example, new renewable energy production is popular. However, it is typically more cost-effective and optimal from a sustainable perspective to improve demand-side energy efficiency. This also makes energy supply available to more customers, which is important in many supply-limited regions of Africa.

One substantial barrier to increased end-use energy efficiency in Africa is the great need for education. It is easier to educate people on the supply-side than on the demand-side. Educating people on the supply-side typically involves one paid person for each 1 MW, while educating industry and business personal in efficient energy end-use involves hundreds to thousands of people per MW capacity. The figure increases to tens of thousands of consumers per MW capacity if the domestic and general public is included. As a result, educating end-users in good energy practice is a major task of national proportions.

Some African countries have endeavored to re-orient its energy-efficiency initiatives to include both supply-side energy production and demand-side end-use. One case-in-point is Ghana.

Ghana submitted a Technology Needs Assessment (TNA) report to United Nations Framework Convention on Climate Change (UNFCCC) in 2003, and received major funding shortly thereafter from the United Nations Development Program (UNDP) and technical support from the National Renewable Energy Laboratory in the U.S. The goal of the TNA is to identify various technology development and transfer programs that can potentially reduce greenhouse gas emissions and contribute to the country's sustainable development.

Top priority end-use technologies identified include replacing incandescent lamps with Compact Fluorescent Lamps (CFLs) and boiler efficiency enhancements. Since the assessment, Ghana has experienced a dramatic increase in the use of CFLs primarily as a result of changes in the country's import tariffs, installation task forces, and sales through retail stores. The CFL promotion policies have been sustainable and self-financing. It is estimated the new CFL market has added U.S. $10 million to the Ghana economy. The CFL deployment has also resulted in a reduction of 6% in Ghana's electricity demand.

Sources:
1. *Energy for Sustainable Development: Energy Policy Options for Africa*, UN-ENERGY/ Africa: 2007.

2.   Stern, N.H., *The Economics of Climate Change: The Stern Review,* Cambridge University Press, Cambridge, UK: 2007.

National Level: Germany

Increased energy efficiency in both energy supply and end-use are critical to the success of Germany's ambitious plan to reduce greenhouse gas emissions by 40% by 2020.[a] Therefore a variety of energy efficiency policies affecting the supply-side and the demand-side are being implemented or will be introduced in the near future. A few examples are provided below.

Energy producers are required to increase efficiency by 3% each year, primarily through the use of modern, highly efficient gas and coal power plants with carbon capture and storage and the increased use of combined heat and power plants. For example, Germany aims to increase the share of combined heat and power in its electricity generation from 12.5% to 25% by 2020.

Starting in 2008, German energy companies will receive 15% fewer emissions allowances to stimulate innovations in energy-efficient power plant technology.

State subsidies are available to anyone that renovates a house or an apartment in an energy-efficient manner. For example, subsidies are available for insulation, replacement of inefficient heating systems, and for new energy-efficient windows.

Energy certificates are going to be required for all new and existing buildings; these certificates will show the building's current energy use and what energy efficiency measures are available.

There will be increased support for a tenant's rights related to energy efficiency. If the landlord of a property does not modernize the property and, as a result, the heating costs are high, the tenant will be allowed a rent reduction.

A motor vehicle tax that is calculated on the basis of CO_2 emissions rather than the size of the vehicle will be introduced soon in Germany. There will also be a motorway toll charge for trucks based on their CO_2 emissions.

Germany currently presides over both the Council of the European Union and the G8. Since combating climate change is of great importance in German energy policy, Germany has made climate and energy central themes of its presidencies of both the European Union and the G8.

---

[a] In comparison, the European Union has committed to a reduction of 30% in greenhouse

gas emissions by 2020 in the case of an international agreement, and to 20% in any event.

Sources:
1. *Taking Action Against Global Warming: An Overview of German Climate Policy,* Federal Ministry for the Environment, Nature Conservation and Nuclear Safety, September 2007.
2. *"Merkel confronts German energy industry with radical policy overhaul,"* Herald Tribune, July 4, 2007.

National Level: China

China's Five Year Plan for 2005-2010 established an ambitious goal of reducing energy intensity, defined as energy use per unit of gross domestic product (GDP), by 20% between 2005 and 2010. This translates into an average reduction of 4% per year. The goal assumes an average annual GDP growth rate of 7.5% from 2005 to 2010; thus energy use can only increase at an annual growth rate of 2.8%. However, both GDP and energy use have been growing faster recently (close to 10% each). As a result, achieving the 20% energy intensity target by 2010 will require a reduction of China's energy use of 19 EJ (or 18 Quads).

To realize the 20% energy intensity reduction goal, China has created the *Top-1000 Energy-Consuming Enterprises* program which sets energy-saving targets for the 1000 largest energy-consuming enterprises in China. These 1000 enterprises used 19.7 EJ in 2004, which represents a third of China's total energy use and close to half of China's industrial energy use. The aggregated goal for the 1000 enterprises is to achieve energy savings of 2.9 EJ (or 2.8 Quads) by 2010.

Steel and chemical industries account for the majority of enterprises included in the program, representing about half of the total. The steel and iron industries also account for ~40% of total energy used by the 1000 enterprises, followed by the petroleum/petrochemical industry (~15%), and the chemical industry (~15%).

Over the summer of 2006, China's National Development and Reform Commission (NRDC) determined energy savings targets for each enterprise. NRDC also held training workshops in the fall of 2006 for all of the enterprises.

Since then, the participants have been requested to conduct energy audits and develop energy actions plans. However, many have found this task difficult because of the lack of qualified energy auditing personnel. To address this barrier, the U.S. Department of Energy (DOE) and NDRC recently signed a memorandum of understanding concerning industrial energy efficiency cooperation. The first phase of this effort involves a team of DOE-assembled industrial energy efficiency experts along with a similar team assembled by NRDC to conduct energy audits jointly at 8-12 enterprises from the "Top 1000" program. As part of the energy audits, DOE energy experts will identify energy-saving opportunities and provide information on U.S. equipment suppliers that can assist in implementing improvements. Additionally, DOE plans to identify potential demonstrations for energy-efficient boilers, fired heaters, and combined heat and power units.

Sources:

1.   Price, L., and W. Xuejun, "Constraining Energy Consumption of China's Largest Industrial Enterprises Through the Top-1000 Energy-Consuming Enterprise Program," *2007 Summer Study on Energy Efficiency in Industry*, White Plains, New York, July 24-27, 2007.

2.   *U.S. and China Sign Agreement to Increase Industrial Energy Efficiency, DOE to Conduct Energy Efficiency Audits on up to 12 Facilities*, Press Release, U.S. Department of Energy, September 14, 2007.

3.   *Memorandum of Understanding between the Department of Energy of the United States of America and the National Development and Reform Commission of the People's Republic of China Concerning Industrial Energy Efficiency Cooperation*, signed in San Francisco, U.S. Department of Energy, September 12, 2007.

National Level: Japan

Japan has been a world leader in implementing energy efficiency measures. As a result, they have one of the lowest levels of greenhouse gas emissions per gross domestic product (GDP) in the world. Japan's 1979 Rational Use of Energy law and its subsequent amendments cover various energy efficiency programs and policies that span the industrial, buildings, and transportation sectors. Representative achievements for each of these sectors are summarized below.

Industry

Energy efficiency achievements in Japan's industrial sector have been significant. Current energy use is at 1970 levels even though the sector has experienced large economic growth. This is due to reductions in the energy intensity of industrial processes, industry restructuring, and energy savings in industrial facilities. Despite improvements, efforts are still underway to reduce the intensity of energy use and greenhouse gas emissions in the industrial sector because it accounts for nearly half of Japan's energy use. Specific energy efficiency efforts include a requirement for medium and large factories to appoint energy conservation administrators, as well as a requirement for them to submit energy conservation plans and reports detailing energy consumption. Another effort relates to a Voluntary Action Plan introduced by the Japan Business Federation, Nippon Keidanren, in 1997. The plan encourages voluntary energy efficiency actions by industry, and has set a goal of reducing CO_2 emissions in the year 2010 to below 1990 levels for targeted industrial businesses.

Buildings

Though the energy efficiency of appliances has steadily improved, energy use in Japan's buildings sector continues to rise due to increased proliferation of electronic devices, growing population, and the desire for more conveniences. To address this problem, Japan has established the Top Runner Program to develop high energy efficiency standards for electrical appliances and devices. This program specifies standards that are equal to or higher than the best available products on the market. For each type of appliance, the program sets a mandatory efficiency requirement for manufacturers and importers to achieve by a specified target year. The standards are continuously reevaluated. As a result of the Top Runner Program, the energy efficiency of many end-use appliances and devices has increased substantially. For example, the efficiency of air conditioners improved by about 40% in 2004 relative to 1997.

In addition, Japan introduced an Energy-Savings Labeling System in August 2000 to inform customers on the energy use characteristics of end-use devices. To help encourage the sales of energy efficient devices, an Energy-Efficient Product Retailer Assessment System was introduced in 2003 to tract and evaluate sales efforts. As of the first of the year, 150 stores were considered active promoters of energy-efficient products. Incentives are also offered to encourage integrated heat and electricity management at plants and office buildings.

To set an example for consumers, Japan established a procurement policy in April 2001 to encourage government entities to purchase energy-efficient end-use equipment for offices and public buildings. Procurement efforts by the government help to create markets for new technologies and to increase market penetration of the products.

Transportation

To reduce greenhouse gas emissions from the transportation sector, Japan has created Top Runner standards for fuel efficiency in passenger vehicles. In addition, the government offers incentives for hybrid vehicles in the form of tax breaks, subsidies, and low-interest loans. The government also offers subsidies for vehicles equipped with an automatic feature to reduce idling. For shippers and large transportation businesses, the government requires that energy-conservation plans and reports be submitted.

Source:
Edahiro, J., "An Overview of Efforts in Japan to Boost Energy Efficiency," *JFS Newsletter*, Sep. 2007.

State Level

California is one of the leading States in terms of energy efficiency in the U.S. There are a large number of effective energy efficiency programs currently in place in California. Many of these programs are implemented by the state's three large investor-owned electric utilities: Pacific Gas & Electric (PG&E), Southern California Edison (SCE), and San Diego Gas & Electric (SDG&E), either together or separately. In addition, municipal utilities are implementing a variety of programs. One representative example of California's achievements is the California Statewide Residential Lighting Program, which was first offered in 2002. The Statewide Residential Lighting Program was designed in response to the 2001 energy crisis experienced in California. Many pilots and full-scale programs were initiated in subsequent years to address capacity limitations, including the Statewide Residential Lighting Program and other energy efficiency and demand response efforts.

The purpose of the 2002 Statewide Residential Lighting Program was to encourage greater penetration of energy efficient lamps and fixtures in the residential sector. The products covered included compact fluorescent lamps (CFLs), torchieres, ceiling fans, and complete fixtures. The program consisted of two types of incentives. One type provided rebates to manufacturers to lower wholesale costs. The other type provided instant rebates to consumers at the point-of-sale.

The program was implemented by PG&E, SCE, and SDG&E. Each utility had in-house management responsibilities. The program leveraged relationships with manufacturers and retailers established in previous lighting programs. Progress was tracked by using data on the number of products delivered by manufacturers and retailer sales information.

In all, 5,502,518 lamps, 24,932 fixtures, 6,736 torchieres, and 50 ceiling fans with bulbs were rebated during the 2002 program year. The total program cost was $9.4 Million. The estimated program accomplishments were 162,888 MWh in energy savings, and 21.4 MW in demand savings. This equates to a reduction in emissions of about 100 thousand metric tons of CO_2 per year. Similar lighting programs have been offered in the years since 2002.

Sources:
1. *National Energy Efficiency Best Practices Study, Vol. R1—Residential Lighting Best Practices Report*, Quantum Consulting Inc., Berkeley, CA: Dec. 2004.
2. Gellings, C.W., and K.E. Parmenter, "Demand-side Management," in *Handbook of Energy Efficiency and Renewable Energy*, edited by F. Kreith and D.Y. Goswami, CRC Press, New York, NY: 2007.

City Level

The city of Portland, Oregon, has been an international leader on community-based energy policy for almost three decades. In 1979, Portland adopted the first local energy plan in the U.S. as a response to the OPEC Oil Embargo. The 1979 energy policy included the establishment of an Energy Office and an Energy Commission. In 1990, Portland adopted a new energy policy that included extensive research and broad community participation of more than 50 public and private groups and associations. The 1990 Energy Policy contained about 90 goals for city operations, energy efficiency, transportation, telecommunications, energy supply, waste reduction, and recycling. The overall goal set in the 1990 Energy Policy is to increase energy efficiency by 10% in all sectors of the city (including the residential, commercial, industrial, and transportation sectors) by 2010.

Portland's successful implementation of the energy policy is primarily a result of the city first focusing on its internal buildings and facilities. Specifically, the city created the City Energy Challenge Program to reduce city energy costs by $1 million by 2000. This goal was achieved, and the energy costs have been reduced further. The energy savings currently equal $2 million per year or more than 15% of the city energy bill.

A few examples of Portland's energy policy successes include:

Energy efficiency improvements in more than 40 million sq.ft. of commercial and institutional space;

Weatherization in more than 22,000 apartment units;

Reduction in per capita household energy use by 9%;

New residential and commercial state energy codes;

Increased bicycle and transit transportation;

Retrofit of all red, green, and flashing amber traffic signals to light emitting diodes (LEDs); and

Use of a wide range of solar-powered equipment and fleet, including water quality monitoring stations, sewer emergency investigation trucks, parking meter repair trucks, and multi-space smart parking meters.

Additionally, on a per-capita basis, Portland's greenhouse gas emissions have fallen by 12.5% since 1993 when Portland became the first U.S. city to adopt a strategy to reduce emissions. Portland's goal is to reduce greenhouse gas emissions to 10% below 1990 levels by 2010.

Source:
Official website of City of Portland, Oregon, http://www.portlandonline.com/.

Corporate Level

In late 2006, Wal-Mart announced its campaign to sell 100 million CFLs by the end of 2007 at its Wal-Mart and Sam's Club stores. Its goal is to sell one CFL to each of its 100 million customers during the year. According to Wal-Mart, if achieved, its goal would save its customers as much as $3 billion in electricity costs over the life of the CFLs. The energy savings would be equivalent to that used by 450,000 single-family homes. About 20 million metric tons of greenhouse gases would be eliminated—comparable to taking 700,000 cars off the road.

Attaining its goal of selling 100 million CFLs by the end of 2007 may not be too difficult. Considering that 90% of Americans live within fifteen miles of a Wal-Mart store and there are 6.6 billion customer visits to Wal-Mart stores each year, all it will take for Wal-Mart to meet its goal is for one customer out of every 60 that enter the store to buy a CFL.

In late 2005, Wal-Mart officials began talking about ways CFLs could help their customers save money. A Wal-Mart staffer asked what difference it would make if the incandescent lamps in all ceiling fans on display in every Wal-Mart store were converted to CFLs. A typical Wal-Mart store has 10 models of ceiling fans on display, each with four lamps. With 3,230 stores and 40 lamps per store, that is nearly 130,000 lamp conversions. At first there was skepticism when it was estimated that changing out those lamps would save nearly $6 million in electric bills annually. But, once the numbers were verified, the decision to go ahead was quickly made.

Wal-Mart's CFL goals are just part of its overall sustainability program. The program goals include:

Being supplied by 100% renewable energy;

Creating zero waste; and

Selling products that sustain the earth's resources and environment.

Towards that end, Wal-Mart opened the first two stores in a series of high-efficiency stores that will use 20% less energy than their typical Supercenter. One store opened on January 19, 2007 in Kansas City, MO, and the other opened on March 14, 2007 in Rockton, IL. Wal-Mart also has two living laboratories (located in Aurora, CO, and McKinney, TX) where they demonstrate and test new energy efficient technologies.

Sources:
1. www.walmartfacts.com, accessed 2007.
2. "Wal-Mart Continues to Change the Retail World—One CFL at a Time," *Power Tools*, Winter 2006-2007, Vol. 4, No. 4, Global Energy Partners, LLC, Lafayette, CA: 2007.

ENERGY EFFICIENCY CHALLENGES IN THE MIDDLE EAST AND NORTH AFRICA

The region encompassing the Middle East and North Africa (MENA) was a focus of the IEA's World Energy Outlook 2005 (International Energy Agency, 2005). This region has a high potential for energy efficiency gains on both the demand-side and the supply-side. A large share of their end-use devices as well as electricity generation capacity is less efficient than in OECD countries.

One of the primary challenges to demand-side energy efficiency is the structure of the energy market. This region is characterized by some of the lowest energy prices in the world. Much of the electricity supplied is heavily subsidized. Electricity rates are particularly under-priced in Iran, Egypt, and countries in the Persian Gulf. In some areas, significant quantities of electricity are lost due to illegal connections despite the low energy costs; in other cases, customers' non-payment of bills is a problem. The low rates present an obstacle to market-driven improvements in end-use energy efficiency. While, in many cases, energy price increases often result in a greater consumer demand for energy-efficient technologies, this price incentive is lacking in the MENA region. Moreover, the approach of raising rates to generate the incentive to purchase energy-efficient technologies (or some type of price reform) would be challenging to implement in the MENA region due to the low income level of many of the customers.

Two of the primary areas of opportunities for end-use energy efficiency improvement are space cooling and desalination. Demands for air conditioning are growing, and the space cooling efficiency in much of the region is below that of the world average. District cooling is one alternative currently receiving significant attention; however penetration is still low (due to the lack of incentive). The region is also home to the largest desalination capacity in the world and the demand for fresh water is expected to triple in some countries such as Saudi Arabia, the UAE, and Kuwait by 2030. Improving the efficiency of desalination processes and combining desalination with power generation are other methods to address energy efficiency. One of the challenges of combined water and power is the mismatch between the electricity demand, which is higher in the summer than in the winter, and the water demand, which is relatively constant.

The power generation efficiency in this region is also on-average significantly lower than in OECD countries. For example, gas-fired plants have average efficiencies of ~33% compared with 43% in OECD and oil-fired plants have average efficiencies of ~34% compared with 42% in OECD. The MENA region often expands capacity with lower-first-cost, less efficient supply-side technologies because of the availability of inexpensive fuel. As a result, opportunities to improve supply-side efficiency are substantial.

There is a growing need for supply-side investments to meet increasing electricity demand, which is projected to increase by 3.4% per year through 2030. Some of this investment could potentially be offset cost-effectively by energy efficiency improvements on both the supply-side and the demand-side. Such improvements would also free up resources for export. The challenge is overcoming the barriers.

CONCLUSION

There is a renewed interest in energy efficiency due to increasing worldwide demand for energy, availability constraints, environmental issues, and economic considerations. Energy efficiency supports sustainable development, energy security, environmental stewardship, and saves money for both energy suppliers and energy end-users. Energy-efficiency measures can begin right away to free up additional supply and to defer the construction of new generation capacity. Energy efficiency is environmentally friendly and cost-effective relative to building new capacity. Much of the technology to enhance energy efficiency is here, yet it is underutilized. Fully implementing all feasible available technology has the potential to make significant improvements in energy efficiency. Nevertheless, additional research and technology innovations will be required to accelerate energy-efficiency improvements in order to meet the expected worldwide growth in energy demand. We already have the experience to develop new technologies and to implement energy-efficiency policies and programs. The oil embargos of the 1970s proved that significant change along these lines is possible with concerted efforts. This book describes a variety of proven tools at our disposal to meet this renewed mandate for energy efficiency, including policies and programs, market implementation methods, and energy-efficient technologies.

References

Realizing the Potential of Energy Efficiency, Targets, Policies, and Measures for G8 Countries, United Nations Foundation, Washington, DC: 2007.

World Energy Outlook 2005: Middle East and North Africa Insights, International Energy Agency, Paris, France: 2005.

Chapter 10

Market Implementation

For voluntary energy efficiency and demand response programs, such as those that might be offered by energy service companies, the success of the program in meeting targets for energy efficiency greatly depends on the level of market penetration achieved. (Note: in this discussion paper, energy efficiency programs are assumed to encompass traditional energy efficiency programs as well as demand response programs.) Planners can select from a variety of methods for influencing consumer adoption and acceptance of voluntary energy efficiency programs. The methods can be broadly classified in six categories. Table 10 1 lists examples for each category of market implementation method. The categories include:

- **Consumer Education:** Many energy suppliers and governments have relied on some form of consumer education to promote general awareness of programs. Brochures, bill inserts, information packets, clearinghouses, educational curricula, and direct mailings are widely used. Consumer education is the most basic of the market implementation methods available and should be used in conjunction with one or more other market implementation method for maximum effectiveness.

- **Direct Consumer Contact:** Direct consumer contact techniques refer to face-to-face communication between the consumer and an energy supplier or government representative to encourage greater consumer acceptance of programs. Energy suppliers have for some time employed marketing and consumer service representatives to provide advice on appliance choice and operation, sizing of heating/cooling systems, lighting design, and even home economics. Direct consumer contact can be accomplished through energy audits, specific program services (e.g., equipment servicing), store fronts where information and devices are displayed, workshops, exhibits, on-site inspection, etc. A major advantage of these methods is that they allow the implementer to obtain feedback from the consumer, thus providing an opportunity to identify and respond to major consumer concerns. They also enable more personalized marketing, and can be useful in communicating interest in and concern for controlling energy costs.

- **Trade Ally Cooperation:** Trade ally cooperation and support can contribute significantly to the success of many energy efficiency programs. A trade ally is defined as any organization that can influence the transactions between the supplier and its consumers or between implementers and consumers. Key trade ally groups include home builders and contractors, local chapters of professional societies, technology/product trade groups, trade associations, and associations representing wholesalers and retailers of appliances and energy-consuming devices. Depending on the type of trade ally organization, a wide range of services are performed, including development of standards and procedures, technology transfer, training, certification, marketing/sales, installation, maintenance, and repair. Generally, if trade ally groups believe that energy efficiency programs will help them (or at least not hinder their business), they will likely support the program.

- **Advertising and Promotion:** Energy suppliers and government energy entities have used a variety of advertising and promotional techniques. Advertising uses various media to communicate a message to consumers in order to inform or persuade them. Advertising media applicable to energy efficiency programs include radio, television, magazines, newspapers, outdoor advertising, and point-of-purchase advertising. Promotion usually includes activities to support advertising, such as press releases, personal selling, displays, demonstrations, coupons, and contest/awards. Some prefer the use of newspapers or the Internet; others have found television advertising to be more effective.

- **Alternative Pricing:** Pricing as a market-influencing factor generally performs three functions: (1) transfers to producers and consumers information regarding the cost or value of products and

Table 10-1. Examples of Market Implementation Methods for Energy Efficiency Programs (Gellings, et al., 2007)

Examples of Market Implementation Methods		
Market Implementation Method	Illustrative Objective	Examples
Consumer education	• Increase consumer awareness of programs • Increase perceived value of energy services	• Bill inserts • Brochures • Information packets • Displays • Clearinghouses • Direct mailings
Direct consumer contact	• Through face-to-face communication, encourage greater consumer acceptance and response to programs	• Energy audits • Direct installation • Store fronts • Workshops/energy clinics • Exhibits/displays • Inspection services
Trade ally cooperation (i.e., architects, engineers, appliance dealers, heating/cooling contractors)	• Increase capability in marketing and implementing programs • Obtain support and technical advice on consumer adoption of energy efficient technologies	• Cooperative advertising and marketing • Training • Certification • Selected product sales/service
Advertising and promotion	• Increase public awareness of new programs • Influence consumer response	• Mass media (radio, TV, Internet, and newspaper) • Point-of-purchase advertising
Alternative pricing	• Provide consumers with pricing signals that reflect real economic costs and encourage the desired market response	• Demand rates • Time-of-use rates • Off-peak rates • Seasonal rates • Inverted rates • Variable levels of service • Promotional rates • Conservation rates
Direct incentives	• Reduce up-front purchase price and risk of energy efficient technologies to the consumer • Increase short-term market penetration • Provide incentives to employees to promote energy efficiency programs	• Low- or no-interest loan • Cash grants • Subsidized installation/modification • Rebates • Buyback programs • Rewards to employees for successful marketing of energy efficiency programs

services being provided, (2) provides incentives to use the most efficient production and consumption methods, and (3) determines who can afford how much of a product. These three functions are closely interrelated. Alternative pricing, through innovative schemes can be an important implementation technique for utilities promoting demand-side options. For example, rate incentives for encouraging specific patterns of utilization of electricity can often be combined with other strategies (e.g., direct incentives) to achieve electric utility demand-side management goals. Pricing structures include time-of-use rates, inverted rates, seasonal rates, variable service levels, promotional rates, off-peak rates, etc. Demand response programs incorporate alternative pricing strategies. A major advantage of alternative pricing programs over some other types of implementation techniques is that the supplier has little or no cash outlay. The consumer receives a financial incentive, but over a period of years, so that the implementer can provide the incentives as it receives the benefits.

- **Direct Incentives:** Direct incentives are used to increase short-term market penetration of a cost control/consumer option by reducing the net cash outlay required for equipment purchase or by reducing the payback period (i.e., increasing the rate of return) to make the investment more attractive. Incentives also reduce consumer resistance to options without proven performance histories or options that involve extensive modifications to the building or the consumer's lifestyle. Direct incentives include cash grants, rebates, buyback programs, billing credits, and low-interest or no-interest loans. One additional type of direct incentive is the offer of free, or very heavily, subsidized, equipment installation or maintenance in exchange for participation. Such arrangements may cost the supplier more than the direct benefits from the energy or demand impact, but can expedite consumer recruitment, and allow the collection of valuable empirical performance data.

Energy suppliers, utilities, and government entities have successfully used many of these marketing strategies. Typically, multiple marketing methods are used to promote energy efficiency programs.

By selecting the appropriate mix of market implementation methods, planners and policy makers can augment or mitigate the external influences, taking into account the customer characteristics, to increase customer acceptance of the demand-side alternative being promoted, thereby obtaining the desired customer response. Figure 10-1 from EOLSS illustrates the customer characteristics, implementation programs, and other external influences that affect three major customer decisions:

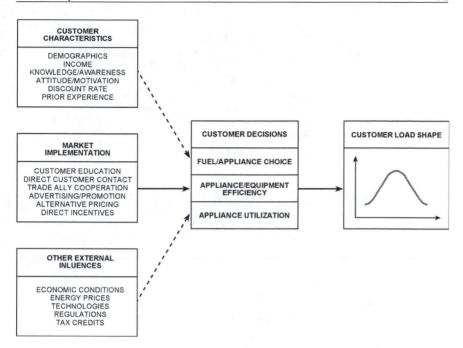

Figure 10-1. Factors Influencing Customer Acceptance and Response The Market Planning Framework

- Fuel/appliance choice
- Appliance/equipment efficiency
- Appliance/equipment utilization

The selection of the appropriate market implementation method should be made in the context of an overall market planning framework. Elements to consider in selecting the appropriate marketing mix, illustrated in Figure 10-2, are:

- Market Segmentation—based on the load shape modification objectives, information on customer end uses and appliance saturation, and other customer characteristics (from consumer research), the market can be broken into smaller homogenous units so that specific customer classes are targeted.

- Technology Evaluation—based on the applicability of available technologies for the relevant end uses and load shape objectives,

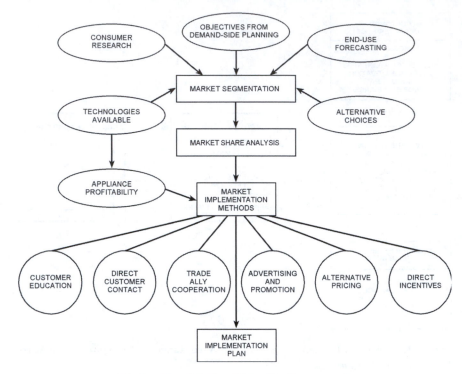

Figure 10-2. Market Planning Framework

the alternative technologies are evaluated and the profitability of specific appliances assessed.

- Market Share Analysis—based on estimates of customer acceptance, the proportion of the total potential market that can be served competitively is estimated.

- Selection of Market Implementation Methods—based on the above analyses and estimates of potential customer acceptance and response, the appropriate mix of implementation method is evaluated and selected.

- Market Implementation Plan—based on the selection of the market implementation methods, an implementation plan is developed to define and execute the demand-side programs.

- Monitoring and Evaluation—the results of implementation are monitored and evaluated to provide relevant information to improve future programs.

The methods used for market segmentation and target marketing can vary, depending on the customer characteristics and the technologies/end uses being addressed by the demand-side alternative. If a technology offers significant benefits to the customer, there is little or no perceived risk, and the customer is aware of the technology and has a favorable attitude toward it, the technology is likely to be well accepted with little need to intervene in the marketplace. However, if the customer acceptance is constrained by one or more barriers, the market implementation methods should be designed to overcome these barriers. Such barriers may include:

- Low return on investment (ROI)
- High first cost
- Lack of knowledge/awareness
- Lack of interest/motivation
- Decrease in comfort/convenience
- Limited product availability
- Perceived risk

FACTORS INFLUENCING CUSTOMER ACCEPTANCE AND RESPONSE

A second important aspect is the stage of "buyer readiness." Customers generally move through various stages toward a purchase decision. The stage that a customer is in will have a bearing on the appropriateness of the market implementation method used. Consumer research can identify where customers are in their decision process. Key questions to consider are:

- Do customers perceive a need to control the cost of energy and are they aware of alternative demand-side technologies?

- Where do customers go to search for more information and guidance on alternatives, and what attributes and benefits are perceived for any given option?

- How much interest is there in participating in a demand-side program, and how can customers be influenced to move toward participation?

- What specific attributes and benefits must customers perceive in order to accept a particular demand-side technology?

- How satisfied are customers who participated in a previous de-mand-side program?

Table 10-2 lists the applicability of Market Implementation Methods to overcome barriers to acceptance. Table 10-3 illustrates applicability to stages of buyer readiness.

Answers to these questions are important in formulating a market implementation program. To influence customer awareness and interest, emphasis can be placed on the use of customer education, direct customer contact, and advertising/promotion. If results of market research indicate that customers in the awareness and interest phase prefer reliability, comfort, and cost-competitive technology options, communicating to customers in advertising/promotion programs should be considered. If the results of consumer research indicate that customers in the purchase/adoption phase use a high implicit discount rate and that first costs are a barrier, the use of direct incentives is appropriate. If first costs are not an obstacle, bit lowering energy bills is a motivating factor, then financial incentives for a particular demand-side technology may be a key consideration.

In addition, if a significant source of information and influence is found to come from trade ally groups during the stage of customer purchase of replacement or new equipment, a point of purchase and cooperative advertising program with trade allies should be considered. Incentive programs for trade allies may also be a consideration.

A final aspect of the buying process is customer satisfaction. As a program is accepted and increases its market share, word of mouth becomes an increasingly important source of customer awareness and interest. Service after the sale is extremely important and represents a form of marketing. Identifying satisfied or unsatisfied customers is useful in terms of evaluating future market implementation methods.

It must be re-emphasized that market segments can be defined using a number of criteria, some of which may be interrelated. Segmenting demand-side markets by end use, stage of buyer readiness, perceived

Table 10-2. Applicability of Market Implementation Methods to Overcome Barriers to Customer Acceptance

Barriers to Customer Acceptance	Customer Education	Direct Customer Contact	Trade Ally Cooperation	Advertising/Promotion	Alternative Pricing	Direct Incentives
Low return on investment (ROI)					H	H
High First Cost, Favorable ROI	M	M		M	M	H
Lack of Knowledge/ Awareness	H	H	M	H		
Lack of Interest/Motivation		M	M	H	M	M
Decrease in Comfort/ Convenience	M	M			M	H
Limited Productivity Availability		M	H			
Perceived Risk	M	H	H		M	H

Blank – Low   M – Medium   H – High

Table 10-3. Applicability of Market Implementation Methods to Stage of Buyer Readiness

Stage of Buyer Readiness	Customer Education	Direct Customer Contact	Trade Ally Cooperation	Advertising/ Promotion	Alternative Pricing	Direct Incentives
Need Recognition Awareness	H	M	M	M		
Search for Alternatives/ Interest	M	H	H	H		
Purchase/Adoption	H	M	H	M	H	H
Satisfaction		M	M			

Blank – Low   M – Medium   H – High

barriers to acceptance, and other socio-demographic factors can suggest the appropriateness of alternative market implementation methods.

Note that the applicability, advantages, and disadvantages of the methods discussed here, as compared to different types of demand-side programs, vary significantly. Typically, planners and policymakers select a mix of the methods most suitable to the relevant demand-side options.

Customer Satisfaction

Many energy suppliers and governments have relied on some form of customer education to promote general customer awareness of programs. Brochures, bill inserts, information packets, clearinghouses, educational curricula, and direct mailings are widely used. Customer education is the most basic of the market implementation methods available and can be used to:

- Inform customers about products/services being offered and their benefits, and influence customer decisions to participate in a program.

- Increase the perceived value of service to the customer.

- Inform customers of the eligibility requirements for program participation.

- Increase the customer's knowledge of factors influencing energy purchase decisions.

- Provide customers other information of general interest.

- Generally improve customer relations.

Whenever a new program is introduced, or changes are made in an existing program, an implementation needs to notify its customers. If a program is new, increasing the general level of customer awareness is an important first step in encouraging market response. Advertising/promotion campaigns can also be used to influence customer awareness and acceptance of demand-side programs (see discussion of advertising and promotion below).

An increasing number of energy suppliers and governments provide booklets or information packets describing available demand-side programs. Detachable forms are frequently provided for customers seeking to request services or obtain more information about a program. Some have also established clearinghouses or toll-free telephone numbers to provide inquiry and referral services to consumers regarding demand-side technologies and programs.

Customer education has the widest applicability to demand-side measures. Some education brochures describe the operation benefits and costs of demand-side technologies (e.g., heat pumps and thermal storage). Others focus on methods to reduce energy costs. Typically, implementers also provide do-it-yourself guides on home energy audits, home weatherization, and meter reading, as well as information on demand-side programs and customer eligibility.

The major advantage of customer education techniques is that they typically provide a more subtle form of marketing, with the potential to influence customer attitudes and purchase decisions. Customer education programs seek to increase customer awareness and interest in demand-side programs. Before customers can decide whether or not to participate in a program, they must have information on the program, its eligibility requirements, benefits, and costs. Customer education techniques should be used in conjunction with one or more other market implementation methods for maximum effectiveness.

Direct Customer Contact

Direct customer contact techniques refer to face-to-face communication between the customer and an energy supplier or government representative to encourage greater customer acceptance of programs. Energy suppliers have for some time employed marketing and customer service representatives to provide advice on appliance choice and operation, sizing of heating/cooling systems, lighting design, and even home economics.

Energy Audits

Are particularly useful for identifying heating/air conditioning system improvements, building envelope improvements, water heating improvements, and the applicability of renewable resource measures. They provide an excellent opportunity for suppliers to interact with customers and sell demand-side options. Energy audits may last from 30 minutes to 3 hours, and cost from $70 to $200 to complete. Energy audits also provide a useful service in obtaining customers feedback and responding to customer concerns.

Program Services

Involve activities undertaken to support specific demand-side measures including heat pumps, weatherization, and renewable energy resources. Examples of such programs include equipment servicing and analyses of customer options. Those can be provided by government employees, energy suppliers, contractors or trade allies.

Store Fronts

A business area where energy information is made available and appliances and devices are displayed to citizens and consumers.

Workshops/Energy Clinics

Special one- or two-day sessions that may cover a variety of topics, including home energy conservation, third-party financing, energy-efficient appliances, and other demand-side technologies.

Exhibits/Displays

Useful for large public showings, including conferences, fairs, or large showrooms. Exhibits can be used to promote greater customer awareness of technologies, appliances, and devices through direct contact. Mobile displays or designed "showcase" buildings can also be used.

Inspection

Typically includes an on-site review of the quality of materials and workmanship associated with the installation of demand-side measures. These inspections are frequently related to compliance with safety or code requirements and manufacturer specifications. Inspections offer the implementer additional opportunities to promote demand-side options.

Direct customer contact methods are applicable to a wide range of demand-side options. A major advantage of these methods is that they allow the implementer to obtain feedback from the consumer, thus providing an opportunity to identify and respond to major customer concerns. They also enable more personalized marketing and can be useful in communicating interest in and concern for controlling energy costs.

Direct customer contact methods are labor-intensive and may require a significant commitment of staff and other resources. Also, specialized training of personnel may be required. Other issues relative to direct customer contact are:

• Scheduling requirements and the need for responding to customer concerns in a timely and effective manner.

- Potential constraints arising from the need for field service personnel to implement direct customer contact programs in addition to either other duties.

- Possible opposition of local contractors/installers to direct installation programs.

- Liability issues related to inspection and installation programs.

- "Fair trade" concerns related to contractor certification for audits or installation.

Trade Ally Cooperation

Trade ally cooperation and support can contribute significantly to the success of many demand-side programs. A trade ally is defined as any organization that can influence the transactions between the supplier and its customers or between implementers and consumers. Key trade ally groups include home builders and contractors, local chapters of professional societies (e.g., the U.S. American Society of Heating, Refrigeration and Air Conditioning Engineers, U.S. American Institute of Architects, The Illuminating Engineering Society, and The Institute of Electrical and Electronic Engineers), technology/product trade groups (e.g., local chapters of the Air Conditioning and Refrigeration Institute and the Association of Home Appliance Manufacturers), trade associations (e.g., local plumbing and electrical contractor associations), and associations representing wholesalers and retailers of appliances and energy consuming devices.

Depending on the type of trade ally organization, a wide range of services are performed, including:

- Development of standards and procedures
- Technology transfer
- Training
- Certification
- Marketing/sales
- Installation, maintenance, and repair.

In performing these diverse services, trade allies may significantly influence the customer's fuel and appliance choice, and/or the appliance or equipment efficiency. Trade allies can, therefore, assist in developing and implementing demand-side programs. For example, trade ally groups can provide valuable information on major technical consider-ations and assist in the development of standards for operating perfor-mance, sizing, and comparison. They can also provide valuable market data on technology sales and shipments. Wholesalers and retailers are particularly valuable in providing cooperative advertising, arranging sales inventory, and supplying information on consumer purchase pat-terns. Installation and service contractors often influence the sale of a demand-side technology, and are responsible for its proper installation and service.

Trade ally groups can be extremely useful in promoting a variety of demand-side measures. Trade ally groups can also help reduce imple-mentation costs. In addition, trade allies can provide technical, logistical, and consumer response information that is useful in the design of utility programs. Generally, if trade ally groups believe that DSM programs will help them (or at least not hinder their business), they will likely support the program.

To obtain the greatest benefit from trade ally cooperation, imple-menters must be willing to compromise and accommodate concerns and questions of allies related to product availability, certification require-ments, paperwork, reimbursement of expenses, cooperative advertis-ing/promotion, training needs, etc. Builders are generally concerned about reducing first costs and, therefore, may resist more expensive building design and appliances. Plumbing and electrical contractors are concerned about the installation and serviceability of heating and cool-ing equipment, and whether an energy supplier promotional campaign will conflict with their peak service periods. Also, manufacturers and retailers prefer to be informed at least six months in advance of an im-plementer's intentions to promote end-use devices or appliances, so that they can have sufficient stocks on hand to meet consumer demand.

Demand-side management programs are less likely to face concerns related to fair trade if the programs are co-sponsored with trade allies. Considerable opposition may result if the program competes directly with trade ally businesses. Implementers will avoid much controversy and possible legal challenges by recruiting trade allies as partners in demand-side programs.

Advertising and Promotion

Energy suppliers and government energy entities have used a vari-ety of advertising and promotional techniques. Advertising uses various media to communicate a message to customers in order to inform or persuade them. Advertising media applicable to demand-side programs include radio, television, magazines, newspapers, outdoor advertising and point-of-purchase advertising. Promotion usually includes activities to support advertising, such as press releases, personal selling, displays, demonstrations, coupons, and contest/awards. Some prefer the use of newspapers based on consumer research that found this medium to be the major source of customer awareness of demand-side programs. Others have found television advertising to be more effective.

Similar to customer education methods, advertising and promotion have widespread applicability. A number of innovative radio and TV spots have been developed to promote demand-side measures. These have included appealing slogans, jingles, and humorous conversations. Other promotional techniques used have been awards, energy-efficient home logos, and residential home energy rating systems, to name a few.

Newspaper Advertising

Newspaper advertising offers a number of advantages, including flexibility in geographical coverage, limited lead time in placing advertisements, intensive coverage within a community, target marketing (by placing ads in certain sections of newspapers), merchandising and advertising design services, and specialized campaigns (e.g., inserts, specialized zone/area coverage for major metropolitan areas, and color preprints).

Disadvantages of newspapers include their relatively short life in the home, hasty reading of articles and ads, and relatively poor reproduction, as compared to magazines.

Specialized regional editions of national magazines and local urban magazines offer the advantages of market sensitivity, relatively longer usage in homes, associated prestige of the magazine, and support services provided to advertisers. A major disadvantage of magazine advertising is the inflexibility compared to spot radio and newspaper advertising. Readership data from magazines can help a utility decide whether magazine advertising is appropriate.

Radio

The use of radio also offers a number of advantages, including frequency of use, lower cost, flexibility in the length and type of communication, audience selectivity, limited advance planning, and mobility. There are also some limitations associated with the use of radio, including fragmented coverage, the fleeting nature of the message, and the limited data on listener characteristics and market share as compared to television advertising.

Television
The mass appeal of television is common knowledge and is illustrated by the large expenditures on television advertising, which reaches a very large audience. Disadvantages of television include the fleeting nature of messages, high production and broadcast costs, limited longevity, and the lack of selectivity.

Outdoor Advertising
There are also various types of outdoor advertising that can be used. Posters are useful in high pedestrian traffic areas or in places where a large audience exists. Smaller posters may be located on buses, in subways, and in transportation stations. Billboards consist of multiple poster sheets and are particularly useful in large open vehicular and pedestrian traffic areas. The advantages of outdoor advertising are the ability to reach large numbers of customers at a low cost. The messages must be simple and easy to comprehend. Rather obvious limitations of outdoor advertising are the brevity of the advertising message, the limited effectiveness, and possible negative aesthetic impacts.

Point of Purchase Promotion
Given the increasing trend toward self-service in retail outlets, point-of-purchase displays are becoming very important. Point-of-purchase displays are oftentimes designed to be consistent with the advertising themes used in other media. Point-of-purchase displays used include appliance stickers; counter top, window, and indoor displays; and models of appliance operation. Point-of-purchase advertising can cover such topics as product features, brand/service loyalty financing, utility marketing programs, and service/maintenance options. The use of point-of-purchase displays in trade ally cooperative advertising should be seriously considered. The major advantage of point-of-purchase advertising is the relatively low cost and the proximity of the advertising message to the point of possible purchase.
Some media may be more useful than others in terms of moving

customers from the "awareness" stage to the "adoption" stage in purchasing a product or service. Therefore, a multi-faceted and carefully scheduled advertising/promotional campaign is worthy of consideration as part of any demand-side management program plan.

An important issue is the extent to which government or state regulatory authorities limit various types of advertising. Advertising and promotion may be limited to those that encourage safety, conservation, load management, etc. Some demand-side management technologies may or may not fall within the definition of allowable advertising. Careful consideration must also be given to the need for statements regarding product liability and warranties (express or implied) in advertising demand-side technologies.

Alternative Pricing

Pricing as a market-influencing factor generally performs three functions:

- Transfers to producers and consumers information regarding the cost or value of products and services being provided.

- Provides incentives to use the most efficient production and consumption methods.

- Determines who can afford how much of a product.

These three functions are closely interrelated. Alternative pricing, through innovative schemes can be an important implementation technique for utilities promoting demand-side options. For example, rate incentives for encouraging specific patterns of utilization of electricity can often be combined with other strategies (e.g., direct incentives) to achieve electric utility demand-side management goals.

Various pricing structures are more or less well-suited to different types of demand-side options. For utilities, time-of-use rates may be generally offered or tied to specific technologies (e.g., storage heating and cooling, off-peaking water heating). They can be useful for thermal storage, energy and demand control, and some efficient equipment options.

For gas utilities, inverted rates used to encourage building envelope and high-efficiency equipment options. Seasonal rates similarly can be used with high-efficiency options.

For both gas and electric utilities, variable service levels can be offered on a voluntary basis as an incentive to reduce demand at certain times of the day. Promotional rates can be used to encourage economic development in an area; and off-peak rates can be particularly applicable to thermal storage and water heating options.

A major advantage of alternative pricing programs over some other types of implementation techniques is that the supplier has little or no cash outlay. The customer receives a financial incentive, but over a period of years, so that the implementer can provide the incentives as it receives the benefits.

One potential disadvantage with some alternative pricing schemes, such as time-of-use rates and rates involving demand charges, is the cost of metering, which can sometimes amount to several hundred dollars per installation. In addition, such rates do not lower the customer's purchase cost, as do some other implementation techniques, such as rebates. Note also that detailed information regarding cost of service and load shapes is needed to design and implement rates to promote demand-side options. Customer education relative to the rate structure and related terminology may also be needed.

Customer acceptance and response to alternative rate structures will vary, depending on the characteristics of the customer, the demand-side programs, and the specific rate structures. For example, research on customer response to electric time-of-use rates indicates that:

- On average, considerable similarity exists in customer response patterns across several, geographically-diverse areas.

- Households typically reduce the share of peak usage in weekday summer use by 3 to 9 percentage points (from 50 to 47-41%) in response to a 4:1 TOU rate.

- The degree of customer response is influenced by appliance ownership and climatic conditions.

Direct Incentives

Direct incentives are used to increase short-term market penetration of a cost-control/customer option by reducing the net cash outlay required for equipment purchase or by reducing the payback period (i.e., increasing the rate of return) to make the investment more attractive. Incentives also reduce customer resistance to options without proven performance histories or options that involve extensive modifications to the building or the customer's lifestyle.

The individual categories of direct incentives include:

- Cash grants
- Rebates
- Buyback programs
- Billing credits
- Low-interest or no-interest loans

Each category is briefly detailed below. While this list is not necessarily all-inclusive (variations and combinations of these incentives are often employed), it indicates the more common forms of direct incentives for large-scale customer adoption. One additional type of direct incentive is the offer of free, or very heavily subsidized, equipment installation or maintenance in exchange for participation. Such arrangements may cost the supplier more than the direct benefits from the energy or demand impact, but can expedite customer recruitment, and allow the collection of valuable empirical performance data.

Cash Grants

These are payments usually one-time sums made to consumers who adopt one or more cost control options. Amounts may be tied to levels of energy or demand reduction or energy efficiency, or can be set simply at a level designed to encourage widespread customer response.

Rebates

Similar to cash grants, rebates are normally single payments made to consumers who install a specific option, either as original equipment or as a replacement for an existing device. Rebate levels are generally set in proportion to the relative benefits of the option to the supplier or implementer. They have been most often used with building envelope options and with efficient equipment and appliance options.

Buyback Programs

These are special incentives that reflect supplier cost savings resulting from the implementations of a mix of cost-control options. The implementer generally estimates the expected average first-year energy use change (and any corresponding demand change) for a particular option through testing or analysis, and then determines its value based on differences between average and marginal costs or other cost criteria. This amount is then normally paid to the customer's installation contractor as a purchase price subsidy. The supplier or implementer in effect "buys back" a portion of the consumer's investment. Buybacks are most often used with building envelope options.

Billing Credits

These are credits applied to a customer's energy bill in return for installing a particular option. Billing credits have most often been used with energy and demand control options, and are generally offered in proportion to the size of the connected load being controlled.

Low-interest or No-interest Loans

These are loans offered at below-market interest rates, or without interest, for the purchase and installation of specific high-efficiency options. They are frequently used to promote the use of high-initial-cost options in the building envelope, efficient equipment and appliance, and thermal storage categories. Reduced interest loans, by allowing home buyers and consumers to finance such expensive items as whole-house insulation, heat pumps, and ceramic storage heaters, often increase the number of consumers willing to invest in these options. Sometimes, low-interest loan programs are co-funded by the energy supplier and government or public agencies.

Direct Incentives and Demand-side Technologies

Direct incentives are being used in a large number of demand-side management programs to encourage customer participation. Various types of direct incentives are applicable to many of the cost control/customer options in each of the major option categories. They can often be used in combination to produce increased customer acceptance.

Rebate and cash grant programs have been praised by some for their administrative simplicity relative to loans. The carrying costs to utilities of low-interest or no-interest loans can be great, as can the recordkeeping requirements. Rebates and grants have an added advantage to the consumer because they significantly lower first costs for new major appliances or other option purchases.

Low-interest loans, however, can often allow consumers to purchase higher-priced options than they would otherwise with only a partial rebate. Thus, if the option produces sufficient benefit, the added costs of the financing program may be justified.

Billing credits can allow the supplier to provide a substantial total incentive to the customer but in small annual amounts over a protracted period. Initial cash outlays are minimal, and administrative costs can be lower than those for loan programs. The consumer must be able to afford the initial purchase, however, and must understand the overall cost-effectiveness of implementing the option.

In developing direct incentive programs, implementers should be aware of potential fair trade or antitrust concerns. If an implementer establishes rational criteria for dealing with trade allies, adheres to those criteria, and uses sound business practices, it is likely that the legality of incentives will not be challenged. If the program is structured so that trade allies support it and also stand to gain from it, then there will probably be no concern about fair trade law violations.

Both customer acceptance and response to direct incentive programs, and potential benefits and costs of these programs, depend on the nature and size of the incentive, the nature of the technology being promoted, and the characteristics of the customer. Experience with direct programs indicates that participants in such programs are generally middle- to upper-class households.

The Market Implementation Plans

Once careful consideration has been given to evaluating alternatives and the appropriate implementation methods have been selected, an implementation plan should be formulated. This plan should contain:

- Program objectives/general purpose—objectives of the market implementation plan must be communicated to all program personnel.

- Measurable program goals (e.g., energy/demand sales or savings and customer contact quotas)—program goals should be well defined and measurable.

- Pre-program logistical planning (i.e., training and equipment/facility procurement)—adequate time must be reserved for employee training, discussions with trade allies, and procurement of the necessary support materials.

- Program implementation procedures—it is a good idea to develop a program implementation flow chart and manual for all key program personnel.

- Management controls—management and program controls must be carefully defined and a customer communication program should be developed and integrated with the demand-side program.

- Program budget/accounting—program cost accounting should be adequately addressed.

- Monitoring and evaluation procedures—develop a program evaluation plan in advance of program implementation that provides sufficiently for program data collection and recordkeeping and offers suitable feedback on program performance.

PROGRAM PLANNING

As with any sophisticated program, a demand-side program should begin with an implementation plan. The plan includes a set of carefully defined, measurable, and obtainable goals. A program logic chart can be used to identify the program implementation process from the point of customer response to program completion. For example, in a direct load control program, decision points, such as meeting customer eligibility requirements, completing credit applications, device installation, and post-inspection can be defined. Actual program implementation can be checked against the plan and major variances reviewed as they occur.

The careful planning that characterizes other energy supply operations should carry over to the implementation of demand-side programs. The programs are expensive and prudent planning will help assure program efficiency and effectiveness. The variety of activities and functional groups involved in implementing demand-side programs further accentuates the need for proper planning.

Program Management

The implementation process involves many different functional entities. Careful management is required to ensure efficient implementation. Managing such widespread activities requires a complete understanding and consensus of program objectives and clear lines of functional authority and accountability.

Ongoing program management is also extremely important. The need for cost accounting, monitoring employee productivity and quality assurance should be addressed; the use of incentives necessitates close monitoring of program costs. For whatever reason, if periodic status reports are required, the exquisite input data and the reporting of key performance indicators must be carefully included.

Program Logistics

Program support includes staffing, equipment, facilities, and training requirements. A program implementation manual is a useful tool to provide program personnel with necessary policy and procedure guidelines. The sample list of functional responsibilities in an implementation program gives an indication of the activities that may be included in such a manual.

A customer adoption plan that coordinates the use of mass media and other advertising and promotional activities (such as bill inserts and direct mail) should be carefully integrated into the implementation program. It is always important to establish good rapport with customers. Consumer concerns should be addressed at all levels of program design and implementation.

Many demand-side management programs include installing a specific piece of equipment or hardware that will alter energy use to benefit both the customer and the supplier.

Some of these technology alternatives can be installed, used or marketed by the energy supplier as part of the demand-side management (DSM) program. The special issues related to such DSM programs involving supplier owned and installed equipment include selecting the proper equipment or hardware, establishing an appropriate customer adoption program, developing quality assurance programs, and developing an installation and maintenance schedule.

Selecting the Proper Equipment or Hardware

Implementers need to evaluate a variety of conflicting factors if they are specifying the functional requirements for equipment or hardware. In the equipment selection process, changes in demand-side technology (such as improvements in the efficiencies of space heating and cooling equipment) and evolving supplier needs (such as automation of the gas or electric utility's distribution system) should be evaluated.

Establishing an Appropriate Implementation Program

Some programs are better suited to promote the installation of certain demand-side technology alternatives. In most cases, a mix of implementation techniques will be used. The selected implementation marketing measures should be compatible with any technology alternative that is part of the demand-side program.

Identifying Quality Assurance Consideration

Because of the possible large number of dispersed devices, implementers can improve customer and utility system performance by considering quality assurance in the implementation program. In the example of electric direct load control program, failures may be attributed either to malfunctions of the devices or to the communication links.

Developing an Installation and Maintenance Schedule

Many expenses are involved in the installation, maintenance, and repair of the numerous devices that are included in a demand-side program. Operating costs can be reduced by developing prudent scheduling policies for limited crew resources. Efficient scheduling of equipment ordering and installation is helpful in reducing unnecessary program delays.

The Implementation Process

Developing, installing and operating an energy supply system that is all the steps associated with "implementing" a supply-side program takes years of planning and scheduling, rigorous analytic modeling, calculations concerning reliability and maintenance, and strict construction scheduling. An equally rigorous approach is needed to implement the demand-side alternatives. There are many actors involved in the implementation process, and this requires the careful coordination of all parties.

The implementation process takes place in several stages. The stages may include forming an implementation project team, completing a pilot experiment and demonstration, and finally, expanding to system-wide implementation. This "time-phased" process tends to reduce the magnitude of the implementation problem, because pilot programs can be used to resolve program problems before system-wide implementation takes place.

Implementing demand-side programs involves many functional elements, and careful coordination is required. As a first step, a high-level, demand-side management project team should be created with representation from the various departments and organizations, and with the overall control and responsibility for the implementation process. It is important for implementers to establish clear directives for the project team, including a written scope of responsibility, project team goals and time frame.

When the limited information is available on prior demand-side program experiences, a pilot experiment may precede the program. Pilot experiments can be a useful interim step toward making a decision to undertake a major program. Pilot experiments may be limited either to a sub-region or to a sample of consumers throughout an area. If the pilot experiment proves cost-effective, then the implementers may consider initiating the full-scale program.

After the pilot experiment is completed, additional effort must be given to refining the training, staffing, marketing, and program administration requirements.

MONITORING AND EVALUATION

Just as there is need to monitor the performance of supply-side alternatives, there is a need to monitor demand-side alternatives. The ultimate goal of the monitoring program is to identify deviations from expected performance and to improve both existing and planned demand-side programs. Monitoring and evaluation programs can also serve as a primary source of information on customer behavior and system impacts, foster advanced planning and organization within a demand-side program, and provide management with the means of examining demand-side programs as they develop.

In monitoring the performance of demand-side management programs, two questions need to be addressed:

- Was the program implemented as planned?
- Did the program achieve its objectives?

The first question may be fairly easy to answer once a routine monitoring system has been adopted. Tracking and review of program costs, customer acceptance, and scheduled milestones can help determine whether the demand-side management program has been implemented as planned.

The second question can be much more difficult to answer. As noted previously, demand-side program objectives can be best characterized in terms of load shape changes. Thus, an assessment of program success must begin with measuring the impact of the program on load shapes. However, this measurement can be difficult because other factors unrelated to the demand-side program can have a significant impact on customer loads.

In monitoring and evaluating demand-side programs, two common approaches can be taken:

- Descriptive—Basic monitoring that includes documentation of program costs, activities completed, services offered, acceptance rates, and characteristics of program participants.

- Experimental—Use of comparisons and control groups to determine relative program effects on participants or nonparticipants, or both.

The two monitoring approaches tend to address different sets of concerns; therefore, it is useful to incorporate both in program design. With a descriptive approach, implementers should be aware of basic program performance indicators, in terms of both administrative procedure and target population characteristics. Information such as the cost per unit of service, the frequency of demand-side equipment installation, the type of participants (single family households or other demographic groups), and the number of customer complaints, can be useful in assessing the relative success of a demand-side management program. Recordkeeping and reporting systems can be helpful in completing descriptive evaluations.

The descriptive evaluation, however, is not adequate for systematically assessing the load shape impacts of the demand-side program. To assess load shape impacts requires the careful definition of a reference baseline against which load shapes with a demand-side management alternative can be judged. The reference baseline reflects those load shape changes that are "naturally occurring"—that is, those changes unrelated to the demand-side program itself.

In some cases, the reference baseline might be the existing forecast with appropriate adjustment to reflect the short-term conditions in the service area. In other cases, the reference might be a control group of customers not participating in the program. The energy consumption and hourly demand of program participants could also be measured before they joined the program to provide a "before" and "after" comparison.

If a "before" reference is used, it is often necessary to adjust data for subsequent changes in the customer's appliance or equipment stock and its usage. Adding an extra appliance (such as a window air conditioner) can more than offset any reductions in energy use resulting from an electric weatherization program. Similarly, the effect of an electric direct load control of a central storage water heater can be altered by changes in a customer's living pattern resulting, for example, from retirement or from the second spouse joining the labor force. The "before" situation must be clearly characterized so that appropriate impact of the program can be measured.

If the reference point is a group of nonparticipants, that group must have characteristics similar to those of the participants. This, of course, requires a great deal of information on both groups to allow for proper matching, including not only appliance stock data but also such information as family size, work schedules, income, and age of head of household.

Monitoring Program Validity

Monitoring programs strive to achieve two types of validity: internal and external. Internal validity is the ability to accurately measure the effect of the demand-side management program on the participant group itself. External validity is the ability to generalize experimental results to the entire population. For example, controlled water heating may reduce peak load for a sample of participants, but there is no guarantee that all customers will react in a similar fashion.

Threats to monitoring program validity usually fall into two categories: problems associated with randomization and problems associates with confounding influences. Randomization refers to the degree to which the participating customer sample truly represents the total customer population involved in the demand-side program. It can also refer to the degree of bias involved in assigning customers to the experimental and control groups.

Confounding influences refer to non-program-related changes that may increase or decrease the impact of a demand-side program. In some cases, the effect of these non-program-related changes can be greater than the effect of the demand-side program. The list of potential sources of confounding influences is extensive, including weather, inflation, changes in personal income, and plant openings and closings.

Data and Information Requirements

Data and information requirements involve the entire process of collecting, managing, validating, and analyzing data in the monitoring and evaluation program. The cost of data collection is likely to be the most expensive part of the evaluation study. Data collection costs can be reduced with proper advance planning and by having sufficient recordkeeping and reporting systems. Sources of evaluation data include program records, energy bills, utility metering, and field surveys. Typically, telephone surveys are to complete field surveys. The data must be valid (measure what it is supposed to measure) and reliable (the same results would occur if repeated, within an acceptable margin of error).

The data collection system should be designed before the implementation of the demand-side program itself. There are a number of reasons for this:

- Some information is needed on a "before" and "after"" program initiation basis. If the "before" data are inadequate, additional data must be collected before the start of the program.

- Some data collections take considerable time, particularly if metering of consumer electric or gas end uses has to be performed.

- Monitoring the effect of the demand-side alternative throughout the program allows for adjustments and modifications to the program.

In addition, information on participant characteristics, awareness of the program, motivations for participating, and satisfaction is very important in evaluating the overall success of a program.

The information-gathering mechanisms may already be in place at many utilities. Load research programs and customer surveys have long been used to collect data for forecasting an planning. The expertise gained in conducting these activities is helpful in considering the development of a monitoring program.

Management Concerns

Monitoring and evaluation programs require careful management attention. Some of the most important hurdles that must be overcome in the management of a monitoring program include:

• Assuring sufficient advanced planning to develop and implement the monitoring and evaluation program in conjunction with the demand-side activity.

• Recognizing that monitoring and evaluation programs can be date-intensive and time-consuming; therefore, evaluation program costs must be kept in balance with benefits.

• Establishing clear lines of responsibility and accountability for program formulation and direction.

• Organizing and reporting the results of the evaluation program to provide management with a clear understanding of these results.

• Developing a strong organizational commitment to adequately plan, coordinate, and fund monitoring programs.

Monitoring and evaluation programs can be organized in four stages: pre-evaluation planning, evaluation design, evaluation design implementation, and program feedback.

References

Decision Focus, Inc., Demand-side Planning Cost/Benefit Analysis, November 1983. Published by the Electric Power Research Institute, Report No. EPRI RDS 94 (RP 1613).

A. Faruqui, P.C. Gupta, and J. Wharton, "Ten Propositions in Modeling Industrial Electricity Demand," in Adela Bolet (ed.) Forecasting U.S. Electricity Demand (Boulder, CO, Westview Press), 1985.

Alliance to Save Energy, Utility Promotion of Investment in Energy Efficiency: Engineering, Legal, and Economic Analyses, August 1983.

Boston Pacific Company, Office Productivity Tools for the Information Economy: Possible Effects on Electricity Consumption, prepared for Electric Power Research Institute, September 1986.

Cambridge Systematics, Inc., Residential End-Use Energy Planning System (REEPS), July 1982. Published by the Electric Power Research Institute, Report No. EA-2512 (RP 1211-2).

"Cooling Commercial Buildings with Off-Peak Power," EPRI Journal, Volume 8, Number 8, October 1983.

Commend building types, the COMMEND Planning System: National and Regional Data and Analysis, EPRI EM-4486.

Customer's Attitudes and Customers' Response to Load Management, Electric Utility Rate Design Study, December 1983. Published by the Electric Power Research Institute, Report No. EPRI SIA82-419-6.

C.W. Gellings and D.R. Limaye, "Market Planning for Electric Utilities," Paper Presented at Energy Technology Conference, Washington, D.C., March 1984.

David C. Hopkins, *The Marketing Plan* (New York: The Conference Board, 1981). Pp. 27-28 and 38-39.

Decision Focus, Inc., Cost/Benefits Analysis of Demand-side Planning Alternatives. Published by Electric Power Research Institute, October 1983, EPRI EURDS 94 (RP 1613).

Decision Focus, Inc. Demand-side Planning Cost/Benefit Analysis, November 1983. Published by the Electric Power Research Institute, Report No. EPRI RDS 94 (RP 1613).

Decision Focus, Inc., Integrated Analysis of Load Shapes and Energy Storage, March 1979. Published by the Electric Power Research Institute. Report No. EA-970 (RP 1108).

Decision Focus, Inc., Load Management Strategy Testing Model. Published by the Electric Power Research Institute, May 1982, EPRI EA 2396 (RP 1485).

Demand-side Management Vol. 1: Overview of Key Issues, EPRI EA/EM-3597, Vol. 1 Project 2381-4 Final Report.

Demand-side Management Vol. 2: Evaluation of Alternatives, EPRI EA/EM-3597, Vol. 2 Project 2381-4 Final Report.

Demand-side Management Vol. 3: Technology Alternatives and Market Implementation Methods, EPRI EA/EM-3597, Vol. 3 Project 2381-4 Final Report.

Demand-side Management Vol. 4: Commercial Markets and Programs, EPRI EA/EM-3597, Vol. 4 Project 2381-4 Final Report.

EBASCO Services, Inc., Survey of Innovative Rate Structures, forthcoming. Electric Power Research Institute, EPRI RP2381-5.

Eco-Energy Associates, Opportunities in Thermal Storage R&D, July 1983. Published by the Electric Power Research Institute, Report No. EM-3159-SR.

Energy Management Associates, Inc., Issues in Implementing a Load Management Program for Direct Load Control, March 1983. Published by the Electric Power Research Institute, Report No. EPRI EA-2904 (RP 2050-8).

Energy Utilization Systems, Inc., 1981 Survey of Utility Load Management Conservation and Solar End-Use Projects, November 1982. Published by Electric Power Research Institute, Report No. EM-2649 (RP 1940-1).

EPRI Project, RP 2547, Consumer Selection of End-Use Devices and Systems.

EPRI Reports prepared by Synergic Resources Corporation. Electric Utility Conservation Programs: Assessment of Implementation Experience (RP 2050-11) and 1983 Survey of Utility End-Use Projects (EPRI Report No. EM 3529).

Gellings, Clark W., Pradeep C. Gupta, and Ahmad Faruqui. Strategic Implications of Demand-side Planning," in James L. Plummer (ed.) Strategic Planning and Management for Electric Utilities, Prentice-Hall, New Jersey, forthcoming 1984.

Identifying Commercial Industrial Market Segments for Utility Demand-side Programs, Gayle Lloyd, Jersey Central Power and Light Company and Todd Davis. Synergic Resources Corporation, The PG and E Energy Expo, April 1966.

J.S. McMenamin and I. Rohmund, *Electricity Use in the Commercial Sector*: Insights from EPRI Research, Electric Power Research Institute Working Paper, March 1986.

Lauritis R. Chirtensen Associates, Inc., Residential Response to Time-of-Use Rates, EPRI Project RP 1956.

Linda Finley, "Load Management Implementation Issues," December, 1982, presentation made at the EPRI Seminar on "Planning and Assessment of Load Management."

M.A. Kuliasha, Utility Controlled Customer Side Thermal Energy Storage Tests: Cool Storage, February 1983. Published by Oak Ridge National Laboratory, Report No. ORNL-5795.

Marketing Demand-side Programs to Improve Load Factor, Electric Power Research Institute, EA-4267, October 1985, p. 5-4.

Mathematical Sciences Northwest, Inc., Reference Manual of Data Sources for Load Forecasting, September 1981. Published by Electric Power Research Institute, EPRI EA-2008 (RP 1478-1).

Methods for Analyzing the Market Penetration of End-Use Technologies: A Guide for Utility Planners, Published by the Electric Power Research Institute, October 1982, EPRI EA-2702 (RP 2045-2).

Michael Porter, *Competitive Strategy* (New York Free Press, 1980).

Non-residential Buildings Energy Consumption Survey: *Characteristics of Commercial Buildings*, 1983. U.S. Department of Energy, Energy Information Administration, p. 57.

Pradeep Gupta. "Load Forecasting," from the Utility Resource Planning Conference sponsored by the University of California-Berkeley, College of Engineering, February 28, 1984. Berkeley, California.

Resource Planning Associates, Inc. Methods for Analyzing the Market Penetration of End-Use Technologies; A Guide for Utility Planners, published by the Electric Power Research Institute, October 1982, EPRI EA-2702 (RP 2045-2). EPRI has also recently funded a project on "Identifying Consumer Research Techniques for Electric Utilities" (RP 1537).

Robert M. Coughlin, "*Understanding Commercial Fuel and Equipment Choice Decisions." Meeting Energy Challenges: The Great PG&E Energy Expo*, 1985. Conference Proceedings, vol. 2, edited by Craig Smith, Todd Davis and Peter Turnbull (New York Pergamon Press, Inc. 1985), pp. 439-447.

Southern California Edison Company, 1981 Conservation and Load Management: Volume II Measurement (1981 Page 2-VIII-I).

State Energy Date Report, U.S. Department of Energy, April, 1986 Statistical Year Book, Edison Electric Institute, 1985.

Stephen Braithwait, Residential Load Forecasting: Integrating End Use and Econometric Methods. Paper presented at Utility Conservation Programs: Planning, Analysis, and Implementation, New Orleans, September 13, 1983.

Survey of Utility Commercial Sector Activities, EPRI EM-4142, July 1985.

Synergic Resources Corporation, Electric Utility Sponsored Conservation Programs: An Assessment of Implementation Mechanisms (forthcoming), Electric Power Research Institute, RP 2050-11.

Synergic Resources Corporation, 1983 Survey of Utility End-Use Projects, Electric Power Research Institute, Report EM-3529-1984 (RP 1940-8).

U.S. Department of Labor, Bureau of Labor Statistics, US DL 85-478, November 7, 1985.

1979 Nonresidential Buildings, Energy Consumption Survey Data for COMMEND buildings.

1983—1987 Research and Development Program Plan published by the Electric Power Research Institute, January 1983, EPRI P-2799-SR.

Gellings, C.W., and K.E. Parmenter, "Demand-side Management," in Handbook of Energy Efficiency and Renewable Energy, edited by F. Kreith and D.Y. Goswami, CRC Press, New York, NY: 2007.

Chapter 11

Efficient Electric End-use Technology Alternatives*

There are numerous electric energy efficient technologies commercially available. Many of these technologies are already in use, but increasing their market penetration has the potential to yield significant improvements in worldwide energy efficiency. In addition, advancements in these technologies as well as commercialization of emerging technologies will act to further energy efficiency improvements. Some advancements will occur naturally, but in order to achieve the maximum potential for energy efficiency, accelerated advancements are needed. Such accelerated advancements require focused research initiatives. Maximization of the energy efficiency potential also means that some of the basic science will need to evolve.

This section lists representative technologies that are commercially available for buildings and industry. It then identifies some emerging technologies as well as some research needs.

There is a second category of electric end-use technologies that is covered in this chapter. It involves those technologies which have the potential to displace end-use applications of fossil fuel while reducing overall energy use and CO_2 emissions. This second group is typically referred to as electrotechnologies.

EXISTING TECHNOLOGIES

Many technologies capable of improving energy efficiency exist today. Some have been established for several decades (e.g., fluorescent

*This chapter benefits from several related efforts, including work from Phase One, Task 2 of the Galvin Electricity Initiative (www.galvinpower.org) and the second includes work done by EPRI staff and Global Energy Partners, LLC on the Impacts of Electrotechnologies—all managed and directed by the author.

lamps), others are new to the market-place (e.g., white LED task lighting), still others have been available for a while, but could still benefit from increased penetration (e.g., lighting controls). Existing end-use technologies which may be deployed in offices and similar commercial applications associated with power plants are also discussed in Chapter 2.

Figure 11-1 lists examples of technologies for buildings and industry. The majority of the technologies listed consume less energy than conventional alternatives. Some of the technologies listed (particularly for industry) are electrotechnology alternatives to thermal equipment. In many cases they are more energy efficient than conventional thermal alternatives; however, in some cases they may use more energy. One of the primary advantages of electrotechnologies is that they avoid on-site emissions of pollutants and, depending on the generation mix of the utility supply and distribution to end-users, they can result in overall emissions reductions.

The buildings technologies are broken down into categories of building shell, cooling, heating, cooling and heating, lighting, water heating, appliances, and general. The industry technologies are divided into the end-use areas of motors, boilers, process heating, waste treatment, air and water treatment, electrolysis, membrane separation, food and agriculture, and general. The table is meant to serve as a representative list of technology alternatives, and should not be considered a complete listing.

Lighting

Artificial illumination is essential to society. It enables productivity, provides safety, enhances beauty, facilitates information transfer, and allows for visual entertainment. The advent of electric lighting drastically transformed modern society's use of artificial illumination. Artificial lighting is now ubiquitous to nearly every aspect of our life. In the U.S., lighting systems currently account for about one-tenth of total electricity consumption in residential buildings, nearly one-quarter of total electricity consumption in commercial buildings, and 6 TO 7% of total electricity consumption in manufacturing facilities. Worldwide, artificial illumination is estimated to demand 20 to 25% of all electric energy in developed countries. In developing nations, artificial illumination is often the first use that newly electrified communities embrace. Because of its significance to society—both in terms

Buildings	Industry
Building shell	**Motors**
Insulation	High efficiency motor
Double- or triple-pane window	Variable speed drive
Low-emissivity window	Multi-speed motor
Window film	**Boilers**
Cooling	Electric boiler
High-efficiency central air conditioner	Firing rate demand controller
Evaporative cooler	Blowdown optimization
High-efficiency room air conditioner	Economizer heat recovery
Displacement ventilation	**Process heating**
Attic venting	Microwave
UV germicidal radiation of chiller coils	Radio frequency
Heating	Induction
Variable-speed furnace	High-temperature plasma arc
Active solar space heater	Infrared
Two stage condensing gas furnace	UV curing
Electrically-heated floor	**Waste treatment**
Zoned resistance heating	Autoclave waste treatment
Cooling and heating	Microwave waste treatment
Smart thermostat	Dry heat waste treatment
Heat pump for space conditioning	Chemical waste treatment
Lighting	Impulse drying
High efficiency fluorescent lamp and ballast	**Air and water treatment**
Compact fluorescent lamp	Ozonation
LED task lighting	UV
LED exit sign	Photocatalytic oxidation
Occupancy sensor	Supercritical water oxidation
Daylighting controls	Non-thermal plasma destruction
Dimming controls	Hybrid processes
Water heating	**Electrolysis**
High-efficiency water heater	Electrocoagulation
Solar water heater	Cerium-catalyzed oxidation process
Heat pump water heater	Electrokinetic remediation of groundwater
Tankless water heater	**Membrane separation**
Heat recovery water heater	Reverse osmosis
Low-flow shower head	Ultrafiltration
Faucet aerator	Microfiltration
Water advanced controls	Geotextile membrane
Water heater thermostat setback	**Food and agriculture**
Water heater cycling	Anaerobic digestion of manure
Water saving appliances: dishwasher, clothes washer	Electric brooder
Appliances	Variable speed vacuum pump
High efficiency refrigerator	Agriculture ventilation fan
High efficiency appliances	Ultra high pressure food processing
Heat pump clothes dryer	Microwave food processing
Duct heat recovery clothes dryer	Irradiation of food products
Induction cooktop	Freeze concentration
Refrigerator or freezer antisweat switch	Ohmic heating
High efficiency electronic devices	Aseptic processing
High efficiency office equipment	Membrane separation
High efficiency pool pump	High efficiency refrigeration
Pool pump timer/ controls	**General**
General	Energy management and controls system
Energy management and controls system	Advanced control and power technologies

Figure 11-1. Examples of Technology Alternatives for Buildings and Industry

of functionality and energy use—maximizing lighting's contribution to a perfect electric energy service system is an obvious goal of future innovation. Table 11-1 shows innovative lighting technologies that may potentially contribute to reduced electricity consumption.

Table 11-1. Innovative Lighting Technologies

Induction Lamps	Full Spectrum/Scotopic Lighting
Pipe Lighting/Solar Light Tubing	White Light-Emitting Diode (LED) Lighting
Multi-Photon-Emitting Phosphor Lighting	Use of Fiber Optics in Light-Sensitive Materials
Concentration Solar	Electroluminescence (EL)
Photoluminescence (PL) Materials	Circadian Lighting/Mood Lighting
Smart Light Combining Fluorescent and LED Technology	Advanced Dimming Using Digital Pulses Over Existing Wiring
Light Engines	

Space Conditioning

Space conditioning is an important consumer function. When properly designed, space-conditioning systems afford the consumer healthy living and working environments that enable productivity and a sense of well-being. Space conditioning is the largest end-user of electricity in both residential and commercial buildings. It also accounts for a relatively large share of electricity use across the manufacturing industries. The primary purposes of space conditioning are to heat, cool, dehumidify, humidify, and provide air mixing and ventilating. To this end, electricity drives devices such as fans, air conditioners, chillers, cooling towers, pumps, humidifiers, dehumidifiers, resistance heaters, heat pumps, electric boilers, and various controls used to operate space-conditioning equipment. Because of its significance and large impact on electricity use across the residential, commercial and industrial sectors, innovations in technologies related to space conditioning may have a substantial effect on how electricity is used in the future. Table 11-2 shows innovative space-conditioning technologies that may potentially contribute to reduced electricity consumption.

Indoor Air Quality

Indoor air quality is a subject of increasing concern to consumers. In fact, the U.S. Environmental Protection Agency (EPA) states that it is one of the five most critical environmental concerns in the U.S. Ac-

Table 11-2. Innovative Space Conditioning Technologies

Electrically Heated Windows	Series Desiccant Wheel for Improved Dehumidification
Electrically Heated Floors	Hydronic Dry Floors
Residential Two-Stage, Condensing Gas Furnaces	Condenser Heat Reactivated Desiccant for Improved Dehumidification
Smart Thermostats	Thermotunneling-Based Cooling
Active Magnetic Regenerative (AMR) Cooling	Ultraviolet Germicidal Irradiation (UVGI) of Chiller Coals
Stirling Engines	Demand-Controlled Hybrid Ventilation
Space Conditioning and/or Water Heating Using Carbon Dioxide (CO_2) Refrigeration Cycle	Heat Pump Water Heating/Space Conditioning

cording to EPA statistics, Americans spend an average of 90% of their time indoors, either at home, school or work. Thus, it is essential for our health and well-being that we take measures to ensure acceptable indoor air quality. The health effects of poor indoor air quality can be particularly severe for children, elderly, and immuno-suppressed or immuno-compromised occupants. Poor indoor air quality is estimated to cause hundreds of thousands of respiratory health problems and thousands of cancer deaths each year (EPA, 2001). Indoor air contaminants such as allergens, microorganisms, and chemicals are also triggers for asthma. In addition to causing illness, poor indoor air quality may inhibit a person's ability to perform, and leads to higher rates of absenteeism. Furthermore, turmoil with the Middle East has promulgated the need for heightened homeland security measures, which focus on protecting building occupants from chemical and biological weapons and creating more resistant indoor environments. Table 11-3 shows innovative indoor air purification technologies that may reduce electricity consumption.

Table 11-3. Innovative Indoor Air Purification Technologies

Photocatalytic Oxidation (PCO) of Air Pollutants	Ultraviolet Germicidal Irradiation (UVGI) for Air Purification
Ion Jet Impact for Air Purification	Air Filtration with Dielectric Barrier Discharge
UVGI Combined with Ozone and Catalytic Oxidation for Air Purification	

Domestic Water Heating

Domestic water heating is essential for the comfort and well-being of consumers. Hot water is used for a variety of daily functions, including bathing, laundry and dishwashing. Water heating is also a significant end user of electricity, particularly for the residential sector. Indeed, water heating accounted for 9.1% of residential electricity use in 2001. It makes up a smaller share in the commercial sector, consuming 1.2% of commercial electricity use in 1999. Electricity is used to run electric resistance water heaters, heat pump water heaters, pumps and emerging devices such as microwave water heaters. Because of the importance of water heating to society—in terms of both functionality and electricity use—maximizing the contributions by water heating technologies to a perfect electric energy service system should be a focus of future innovation. Table 11-4 shows innovative domestic water heating technologies which may potential contribute to reduced electricity consumption.

Table 11-4. Innovative Domestic Water Heating Technologies

Microwave Water Heating and Purification	Concentration Solar
Space Conditioning and/or Water Heating Using Carbon Dioxide (CO_2) Refrigeration Cycle	Heat Pump Water Heating/Space Conditioning

Hyper-efficient Appliances

While the adoption of the best available energy-efficient technologies by all consumers in 100% of applications is far from complete, the utilities, states and other organizations interested in promoting increased efficiency are beginning to question the availability of the "next generation" of energy-efficient technologies. After several years of unsuccessfully trying to encourage the electricity sector to fill in the gaps of R&D in advanced utilization appliances and devices, EPRI has turned instead to the goal of transferring proven technologies from overseas.

As a result of several factors, manufacturers of electrical apparatus in Japan, Korea and Europe have outpaced U.S. firms in the development of high-efficiency electric end-use technologies. If fully deployed, these technologies could reduce the demand for electric energy by over 10%. In addition, collectively these technologies have the potential to reduced electric energy consumption in residential and commercial applications by up to 40% for each application. They represent the single greatest opportunity to meet consumer demand for electricity.

The technologies are currently being demonstrated with several utilities in different climate regions to assess their performance when deployed in diverse environments. This will ensure a thorough evaluation. The following technologies are considered among those ready for demonstration in the U.S.:

- Variable Refrigerant Flow Air Conditioning (with and without ice storage)
- Heat Pump Water Heating
- Ductless Residential Heat Pumps and Air Conditioners
- Hyper-Efficient Residential Appliances
- Data Center Energy Efficiency
- Light-emitting Diode (LED) Street and Area Lighting

Ductless Residential Heat Pumps and Air Conditioners

Approximately 28% of residential electric energy use can be attributed to space conditioning. Use of variable frequency drive air conditioning systems can offer a substantial improvement when compared to conventional systems.

In addition, in many climate zones, the industry has long recognized that the application of electric-driven heat pump technology would offer far greater energy effectiveness than fossil fuel applications. However, except in warmer climates, the cost and performance of today's technology in insufficient to realize that promise. These ductless systems have the potential to substantially change the cost and performance profile of heat pumps in the U.S.

Variable Refrigerant Flow Air Conditionings

Ducted air conditioning systems with fixed-speed motors have been the most popular system for climate control in multi-zone commercial building applications in North America. These systems require significant electricity to operate and offer no opportunity to manage peak demand.

Multi-split heat pumps have evolved from a technology suitable for residential and light commercial buildings to variable refrigerant flow (VRF) systems that can provide efficient space conditioning for large commercial buildings. VRF systems are enhanced versions of ductless multi-split systems, permitting more indoor units to be connected to each outdoor unit and providing additional features such as simultaneous heating and cooling and heat recovery. VRF systems are very popular in Asia and Europe and, with an increasing support available from major U.S. and Asian manufacturers, are worth considering for multi-zone commercial building applications in the U.S.

VRF technology uses smart integrated controls, variable-speed drives, refrigerant piping and heat recovery to provide products with attributes that include high energy efficiency, flexible operation, ease of installation, low noise, zone control and comfort using all-electricity technology.

Ductless space conditioning products, the forerunner of multi-split and VRF systems, were first introduced to Japan and elsewhere in the 1950s as split systems with single indoor units and outdoor units. These ductless products were designed as quieter, more efficient alternatives to window units (Smith, 2007).

Heat Pump Water Heating

Heat pump water heaters (HPWHs) based on current Japanese technology are three times more efficient than electric resistance water heaters and have the potential to deliver nearly five times the amount of hot water, even compared to a resistance water heater.

HPWHs are significantly more energy efficient than electric resistance water heaters, and can result in lower annual water heating bills for the consumer, as well as reductions in greenhouse gas emissions. But the high first costs of heat pump water heaters and past application and servicing problems have limited their use in the U.S.

Water heating constitutes a substantial portion of residential energy consumption. In 1999, 120,682 GWh of electricity and 1,456 trillion Btu of natural gas were consumed to heat water in residences, amounting to 10% of residential electricity consumption and 30% of residential natural gas consumption (EPRI, 2001). While both natural gas and electricity are used to heat water, the favorable economics of natural gas water heaters have historically made them more popular than electric water heaters.

Heat pump water heaters, which use electricity to power a vapor-compression cycle to draw heat from the surrounding environment, can heat water more efficiently for the end user than conventional water heaters (both natural gas and resistant element electric). Such devices offer consumers a more cost-effective and energy-efficient method of electrically heating water. The potential savings in terms of carbon emissions at the power plant are also significant. Replacing 1.5 million electric resistance heaters with heap pump water heaters would reduce carbon emissions by an amount roughly equivalent to the annual carbon emissions produced by a 250 MW coal power plant.

Heat pump water heaters have been commercially available since the early 1980s and have made some inroads in some places in the world, particularly in Europe and Japan.

Hyper-efficient Residential Appliances

Driven in part by high electricity prices and government encouragement, Japanese, Korean, Vietnamese and European markets have witnessed the introduction and widespread adoption of "hyper-efficient" residential appliances including electric heat pump clothes washers and dryers, inverter-driven clothes washers, multi-stage inverter-driven refrigerators, and advanced-induction ranges and cook tops.

Depending on the application, these appliances can use 50% less electricity than conventional U.S. appliances. However, there are issues with regard to their acceptance with U.S. consumers and their actual performance.

Data Center Energy Efficiency

Data centers consume 30 terawatt hours of electricity per year. The technologies that are employed in those buildings today only allow 100 watts of every 245 watts of electricity delivered to actually be used to provide computational ability. The steps in between delivery to the building and actual use include the following conversions:

- Uninterruptable Power Supplies   88—92% Efficient
- Power Distribution   98—99% Efficient
- Power Supplies   68—75% Efficient
- DC to DC Conversion   78—85% Efficient

In addition, all the lost energy has to be cooled. That is typically done with air conditioning requiring 1,000 watts for each ton of cooling, typically at an efficiency of 76%.

Light-emitting Diode (LED) Street and Area Lighting

Street lighting is an important lifestyle enhancement feature in communities all over the world. There is a move across the U.S. to replace existing street and area lights—normally mercury vapor, high-pressure sodium (HPS) or metal halide (MH) lamps—with new technology that costs less to operate, and LEDs are at the forefront of this trend. Since LED street and area lighting (LEDSAL) technology is still relatively new to the market, utilities, municipalities, energy service providers and light designers have expressed a keen interest in what the tradeoffs are between conventional lighting and LEDSAL. Cost is probably first among them, with the disadvantage of higher initial cost, but the advantage of lower operating costs.

Several important tradeoffs to consider when adopting LEDSAL are presented in this chapter, organized according to their advantages and disadvantages. The advantages include energy efficiency, lower operating costs, durability, flexibility and improved illumination that can lead to increased safety. On the disadvantages side are higher first costs, lower immunity to electrical disturbances, lower LED efficacy, varying fixture designs, three-wire installation, and unsuitability for retrofits into conventional fixtures.

LEDSALs offer a number of advantages related to power and energy use, light quality, safety and operating costs.

INDUSTRIAL

Industrial use of electricity is dominated by the use of electric motors, lighting and process heating. Lighting applications have been discussed earlier. This section focuses on motors and drives as well as process heating. This is followed by three additional key opportunities for energy efficiency in the industrial sector—cogeneration, thermal energy storage and the application of industrial energy management programs.

Motors and Drives

Electric motors and drives use about 55% of all electricity in the

U.S. In addition, electrically driven equipment accounts for about 67% of industrial electricity use in the U.S. There are several types of motors used in industrial applications, including DC motors, permanent magnet DC motors, synchronous motors, reluctance motors and induction motors. Induction motors are by far the most common motor used in industry and account for over 90% of all the motors of 5 horsepower and greater. As a result of their prevalence, the efficient use of motors and drives presents a considerable opportunity for energy savings in the industrial sector and beyond. Applications of electric drives include compressors, refrigeration systems, fans, blowers, pumps, conveyors, and assorted equipment for crushing, grinding, stamping, trimming, mixing, cutting and milling operations.

It is best to focus on the entire drive system to realize maximum energy savings. A drive system includes the following components: electrical supply, electric drive, control packages, motor, couplers, belts, chains, gear drives and bearings. There are losses in each component that need to be addressed for maximum efficiency. In general, efficiency improvements can be made in four main categories: the prime mover (motor), drive controls, drive train, and electrical supply. Indirect energy savings can also be realized through efficient motor and drive operation. For example, less waste heat is generated by an efficient system, and therefore, a smaller cooling load would result for an environment that is air conditioned. This section focuses on efficiency opportunities in (a) the operation and maintenance of electric drive systems, (b) equipment retrofit and replacement, and (c) controls and alterations to fans, blowers, and pumps.

The efficiency of motors and drives can be improved to some extent by better operation and maintenance practices. Operation and maintenance measures are typically inexpensive and easy to implement, and provide an opportunity for almost immediate energy savings. The main energy efficiency opportunities for motors, drive trains, and electrical supply systems are described below.

Motors
- **Better lubrication:** It is important to use high-quality lubricants that are appropriate for the particular application. Too much or too little lubrication can reduce system efficiency.

- **Improved cooling:** Adequate cooling of motors can reduce the need for motor rewinds and improve efficiency. Cleaning heat transfer surfaces and vents will help improve cooling.

- **Spillage prevention:** Prevent the spillage of water into motor windings. This is facilitated by choosing leak-proof motors

- **Minimized low- or no-load operation of motors:** Eliminate motors that are operating infrequently or at low loads. Efficiency decreases as the percentage of loading is decreased.

- **Motor matching:** Size the motor correctly to fit the application. Match its torque characteristics to the load.

- **Quality rewinding:** Use high-quality winding techniques and materials (such as copper) when rewinding motors. Consider replacing or rewinding motors with aluminum windings.

- **Operation:** Analyze the merits of continuous vs. batch operation. Is it more efficient to run smaller motors continuously, or larger motors in batch operation?

Drive Train
- **Belt operation:** Properly align belt drives. Adjust belt tension correctly.

- **Minimization of friction:** Reduce losses due to friction by checking operation of bearings, gears, belt drives and clutches.

- **Lubrication:** Lubricate the chain in chain drives correctly. Use appropriate types and quantities of lubricants.

- **Synchronous belts:** Convert V-belts to synchronous belts.

- **Chains:** Convert roller chains to silent chains.

- **Direct driven loads:** Use direct-driven loads in the place of gear, belt or chain drives for maximum efficiency.

- **Quality bearings:** Use high-quality bearings for minimized friction.

Electrical Supply
- **Operation at rated voltage:** Motors are most efficient if they are operated at their rated voltage.

- **Phase balance:** Balance three phase power supplies.

- **Efficient power systems:** Losses can occur in the power systems that supply electricity to the motors. Check substations, transformers, switching gear, distribution systems, feeders and panels for efficient operation. De-energize excess transformer capacity.

Equipment Retrofit and Replacement
Equipment retrofit and replacement measures require more money and time to implement than do operation and maintenance measures; however, they can also result in more significant energy savings. Some common retrofit and replacement opportunities for motors and drives are described below.

- **Heat recovery**: Modify equipment to recover heat. The waste heat can supply heat for another part of the process, reducing the demand on heating equipment.

- **Controls for scheduling:** Install controls to schedule equipment. Turn off motors when they are not in use, and schedule large motors to operate during off-peak hours.

- **Other controls:** Consider power factor controllers in low-duty-factor applications, and feedback control systems.

- **Variable speed drives (adjustable speed drives):** Install variable speed drives to control the shaft speed of the motor. This reduces energy consumption considerably by matching the motor speed to the process requirements.

- **Replacement of throttling valve with variable-speed drive:** Control shaft speed with a variable-speed drive instead of a throttling valve. Throttling valves are associated with significant energy losses.

- **Replacement of pneumatic drives:** Consider replacing pneumatic drives with electric motors, if possible. Pneumatic drives use electricity to generate compressed air which then is converted to mechanical energy. Electric motors are much more efficient; but in some applications, pneumatic drives are preferred because of electrical hazards or because of the need for lightweight and high-power drives. The main inefficiency of pneumatic drives arises from air leaks, which are hard to eliminate or avoid.

- **Replacement of steam jets:** Replace steam jets on vacuum systems with electric motor-driven vacuum pumps.

- **High-efficiency motors:** Install high-efficiency motors in all new designs and system retrofits, and when motors need replacement. Motor manufacturers have focused on improving motor efficiency since the mid-1970s when the cost of electricity started to rise. Despite motor innovations and availability, the utilization of high-efficiency motors in industry is small compared to what is possible; there is still a large potential for energy savings.

Process Heating

Process heat accounts for 10% of industrial end-use of electricity in the U.S. Although this percentage is small compared to electric drive systems, it is significant enough that energy efficiency improvements in process heat applications have the potential for a substantial impact on overall electrical efficiency. The four main ways for process heat to be generated are with combustible fuel-based systems, electric-based systems, thermal recovery systems, or with solar collection systems. When all types of process heat are considered, electrically powered systems only account for a few percent of the total. The share of electric process heat systems is likely to increase in the future because of several advantages associated with electric systems, including ease of control, cleanliness at the point of use, safety, small size, and applicability for a large range of capacities. In addition, electric systems will likely become more prevalent as a result of the increasing costs of combustible fuels.

Generally, process heat is used for melting, heating and drying operations. Specific applications include distilling, annealing, fusing, cooking, softening and moisture removal. There are a variety of electric heating systems currently available, and there are many emerging technologies. The most common electric process heat technologies include resistance heaters, induction heaters, infrared systems, dielectric systems (RF and microwave), electric salt bath furnaces, and direct arc electric furnaces. The following section of this chapter summarizes some of the energy efficiency opportunities for process heat applications.

Cogeneration

In the early 1900s, most industries generated their own power and used the waste heat for supplemental thermal energy. This was the beginning of cogeneration. As the public utility industry grew in size, and the costs of electricity generation decreased, cogeneration use at the industry level started to decline. Between 1954 and 1976, industrial electrical power production decreased from about 25 to 9% and continued to decline. However, as a result of the oil embargo in the 1970s, cogeneration began to receive more notice and has experienced a slow growth since the mid-1980s when industrial cogeneration capacity was about 4%. In the early 1990s, it increased to about 5.1%. Cogeneration should continue to grow in the face of increased fuel prices and the development of new cost-effective technologies.

Cogeneration systems produce mechanical energy and thermal energy, either simultaneously or sequentially. Electricity is then produced when the mechanical energy is applied to a generator. There are three main classes of cogeneration systems: (1) topping cycles, (2) bottoming cycles, and (3) combined cycles. In topping cycles, electricity is produced first, and the thermal energy that results from the combustion process is used for process heat, space heat or additional electricity production. In bottoming cycles, high-temperature thermal energy (typically steam) is produced for process applications, and then the lower temperature steam is recovered and used to generate electricity. Topping cycles are more commonly used than bottoming cycles. Combined cycles are based on topping cycles, but the steam that is generated from the exhaust gas is directed into a steam turbine to produce additional electricity.

Cogeneration systems are currently in use in a variety of industries, and the number of candidate industries is increasing. Any industry that has a demand for thermal energy and electrical energy is a candidate for cogeneration. Some of the main applications of cogeneration in the industrial sector include:

- **Pulp and paper industry:** The pulp and paper industry has been the primary industrial user of cogeneration. Large quantities of burnable wastes are used to fuel cogeneration systems for electricity generation.

- **Chemical industry:** The chemical industry is another significant user of cogeneration. Historically, the chemical industry has had the third largest installed cogeneration capacity in the industrial sector. It uses about the same quantity of steam annually as the pulp and paper industry. The large thermal demand makes cogeneration a desirable option.

- **Steel industry:** Open-hearth steel-making processes produce an off-gas that is capable of providing fuel to produce steam. The steam is then used to drive blast furnace air compressors and for other applications. Cogeneration is applicable (and currently used) for steel mills of the open-hearth type; however, newer mills utilizing electric-arc technology do not have a significant thermal demand, and therefore, cogeneration is not as applicable.

- **Petroleum refining industry:** The petroleum refining industry has a significant thermal demand and is very well suited for cogeneration.

- **Food processing:** The demand for cogeneration in the food industry is on the rise. The main limitation is the cyclic nature of its thermal demand.

Thermal Energy Storage

Thermal energy storage (TES) can also be considered an industrial energy source. TES is the storage of thermal energy in a medium (i.e., steam, water, oil or solids) for use at a future time. TES is used to manage energy in several ways. For example, TES provides peak load coverage for variable electricity or heat demands, it allows for the equalization of heat supplied from batch processes, and it enables the storage of energy produced during off-peak periods for use during peak periods. TES is also used to decouple the generation of electricity and heat in cogeneration systems. Better energy management through the use of TES can help industries reduce their dependency on utilities for energy supply. In addition, TES systems can provide a backup reserve of energy in the event of a power outage.

Industrial Energy Management Programs

The establishment of an energy management program is a crucial part of the process of setting and achieving industrial energy-efficiency goals. First and foremost, the establishment of an energy management program requires a commitment from management to initiate and support such a program. Once management is committed, an energy management program should be custom designed for each specific application, since efficiency goals vary with the type and size of the industry. However, there are several main guidelines that are applicable to any energy management program. In general, the procedure for setting up an energy management program requires the following six main steps:

- Appoint energy managers and steering committee
- Gather and review historical energy use data
- Conduct energy audits
- Identify energy-efficiency opportunities
- Implement cost-effective changes
- Monitor the results

This general procedure may be applied to any type of facility, including educational institutions, commercial buildings, and industrial plants.

Manufacturing Processes

The industrial revolution brought about radical changes in how items were produced. The automation of manufacturing processes has improved the modern world's standard of living and continues to do so. In recent times, these productivity gains have come from the introduction of the computer into manufacturing, automating repetitive tasks and allowing for improved quality control and process management. Manufacturing processes today are heavily automated and dependent on robots and computers to perform functions. This degree of automation, in turn, requires reliable and high-quality power. Manufacturing is also very energy intensive and can have negative environmental impacts. Technological innovations that can improve upon existing manufacturing processes, making them more productive, cost-effective, energy-efficient, and environmentally responsible are depicted in Table 11-5.

Table 11-5. Innovative Technologies for Manufacturing and Control of Air Emissions

Electron Beam Irradiation	Power Electronics
DC-Arc Plasma Furnace for Melting Metals	Laser Cutting Tools
Powder Metallurgy (PM)	Three-Dimensional (3-D) Printing
Advanced Oxidation Processes	Biological Processes for Air Treatment
Variable-Speed Motor Drives	

ELECTROTECHNOLOGIES

Electrotechnologies, while often existing technologies, are most often categories of end-use technology not considered as energy-efficient. Interestingly, there are many end uses of fossil fuels that are inefficient from a total energy balance and environmental perspective. This is due to the physics of energy conversion wherein many end-use applications of electricity are far superior in conversion of electricity to actual desired heat, motive power, comfort or other derived energy conversion need that fossil fuels are. These electrotechnologies save so much energy at the point of end use so as to more than offset losses in electricity production and delivery. Also, as a result, they have lower CO_2 emissions.

Residential Sector

There are eight electrotechnologies typically considered as having the potential for consideration as superior when compared to fossil-fuel end-use applications in the residential sector. Table 11-6 summarizes these eight electrotechnologies.

Table 11-6. Residential Electrotechnologies

Heat Pump Clothes Dryer	Electric Convection Oven
Electric Induction Range Top	Heat Pump Pool/Spa Heater
Air-Source Heat Pump, Cooling	Air-Source Heat Pump, Heating
Ground-Source Heat Pump, Cooling	Ground-Source Heat Pump, Heating
Electric Instantaneous Water Heater	Heat Pump Water Heater

Commercial Sector

Twenty electrotechnologies have the potential for consideration as electrotechnologies in the commercial sector. Table 11-7 lists the twenty electrotechnologies in the commercial sector.

Industrial Sector

Table 11-8 summarizes the industrial electrotechnologies typically considered as having the potential to be superior when compared to

Table 11-7. Commercial Electrotechnologies

Heat Pump Clothes Dryers	Electric Braising Pan
Electric Broiler	Electric Griddle
Electric Fryer, Flat Bottom	Electric Fryer, Open Deep Fat
Electric Fryer, Pressure/Kettle	Electric Oven, Conveyor
Electric Oven, Deck	Electric Oven, Rotisserie
Electric Oven, Standard/Convection/Combination	Electric Range Top
Electric Steamer, Compartment	Electric Steamer, Kettle
Electric Wok	Heat Pump Pool/Spa Heater
Electric Boilers	Air-Source Heat Pump, Cooling
Air-Source Heat Pump, Heating	Ground-Source Heat Pump, Cooling
Ground-Source Heat Pump, Heating	Heat Pump Water Heater

Table 11-8. Industrial Electrotechnologies

Electric Boilers	Electric Drives
Heat Pumps	Induction Heating
Radio Frequency Heating	Microwave Heating
Electric Infrared Heating	UV Heating
Electric Arc Furnace	Electric Induction Melting
Plasma Melting	Electrolytic Reduction

fossil-fuel applications. A few are described in more detail in the section which follows.

Induction Process Heating

Induction heating systems use electromagnetic energy to induce electric currents to flow in appropriately conductive materials. The materials are then heated by the dissipation of power in the interior of the materials. Induction heating is similar to microwave heating except that induction heating utilizes lower frequency, longer wavelength energy. The approximate frequency range is 500 to 800 kHz.

Dielectric Process Heat

Dielectric heating is accomplished with the application of electromagnetic fields. The material is placed between two electrodes that are connected to a high-frequency generator. The electromagnetic fields excite the molecular makeup of material, thereby generating heat within the material. Dielectric systems can be divided into two types: RF (radio frequency) and microwave. RF systems operate in the 1 to 100 MHz range, and microwave systems operate in the 100 to 10 000 MHz range. RF systems are less expensive and are capable of larger penetration depths because of their lower frequencies and longer wavelengths than microwave systems, but they are not as well suited for materials or products with irregular shapes. Both types of dielectric processes are good for applications in which the surface to volume ratio is small. In these cases, heating processes that rely on conductive, radiative and convective heat transfer are less efficient.

Infrared Process Heat

Infrared heating is used in many drying and surface processes. It is based on electromagnetic radiation at small wavelengths of one to six microns and high frequencies (~108 MHz). Because of their small wavelengths of energy, infrared heating systems typically are not capable of penetrating more than several millimeters into materials and are, therefore, best suited for surface applications. Typical applications include drying paper and textiles, hardening surface coatings, and accelerating chemical reactions.

Electric Arc Furnaces

Electric arc furnaces use a large percentage of the energy that is consumed in the primary metals industry. They are used primarily for melting and processing recycled scrap steel. The refinement of scrap steel requires only about 40% of the energy required to produce steel from iron ore in a typical oxygen furnace. Since its inception, electric arc technology has improved substantially, and improvements are continuing. The percentage of steel produced in arc furnaces increased from 15% in 1970 to 38% in 1987, and the percentage is still rising. In addition, the electricity use per unit of product has decreased significantly. Energy improvements include the use of better controls, preheating of the material with fuel combustion or heat recovery, waste minimization by particle recovery, and ladle refining.

Efficiency Advantages of Electric Process Heat Systems
- **Quick start-up:** Electromagnetic systems are capable of quick start-up. Fuel-fired furnaces require long warm-up periods. As a result equipment is often left on continuously. Electric systems can be turned off when not in use.

- **Faster turn-around:** Electric systems accomplish the required heating at a faster rate than furnaces. This can result in increased productivity and smaller heating times.

- **Less material loss:** Faster electric heating rates result in less material scaling. The amount of scaling is related to the time and quantity of exposure to oxygen at high temperatures. Energy is indirectly saved in the form of less material loss.

- **Direct heating process:** Direct heating systems are more efficient than indirect systems because energy losses from the heat containment system to the work piece are eliminated. Implementation of direct electric heating with infrared or dielectric technology can reduce energy use for industrial process heating by up to 80%, with typically a payback period of one to three years.

- **Heat generation inside material:** In induction and dielectric heating systems, heat is generated throughout the material, regardless of the material's thermal conductivity. This is in contrast to radiative and convective heating systems, in which the effectiveness and efficiency depends on the material's thermal conductivity. Since the materials are heated from within, less energy is lost and the heat is distributed more uniformly. This also results in increased product quality.

- **More process control:** By varying the frequency of the energy, the heating parameters can be optimized for the specific application. As a result, energy loss is reduced. In addition, electric processes can direct the energy to the desired location more precisely. Energy use is reduced by the avoidance of heating unnecessary material or equipment.

MERITS OF ELECTROTECHNOLOGIES BEYOND ENERGY EFFICIENCY

There are several benefits to the increased use of electrotechnologies in the residential sector aside from the energy and CO_2 savings that they offer. A few of the specific merits of the electrotechnologies selected for analysis relative to their fossil-fueled counterparts include the following:

- **Urban Emissions Reduction:** Emissions from direct combustion of fossil fuels are moved from homes to central generation sites, which tend to be located farther from city centers.

- **Heat Pumps Leverage Ambient Heat:** Therefore, in many heating applications, the primary energy source is the sun, which is a renewable and local energy source.

- **Dehumidification:** Heat pump water heaters cool and dehumidify the surrounding air when operating.

- **Manufacturing Development:** Wider adoption of heat pump technologies in the U.S. will present the opportunity for manufacturing development in the U.S. and, consequently, job creation in the green manufacturing sector.

One unique characteristic of electricity is that it is the only fuel source that can decrease its CO_2 intensity. On one hand, this is due to the ability to change the fuel mix of generation, i.e., increased penetration of nuclear and renewable generation sources. On the other hand, there is increased research and development relating to technologies to reduce the amount of CO_2 entering the atmosphere at fossil-fueled generation sites, i.e., carbon capture and storage. In contrast, direct combustion of fossil fuels will always produce the same amount of CO_2, with only small variations based on fuel composition.

In the industrial sector, electrotechnologies have some unique additional advantages over fossil-fueled technologies in industrial processes. For one, electricity is an orderly energy form, in contrast to thermal energy, which is random. This orderliness means that electrical processes are controllable to a much more precise degree than thermal processes. In addition, since electricity has no inertia, energy input can instantly adjust to varying process conditions such as material temperature, moisture content, or chemical composition. For instance, lasers and electron beams can produce energy densities at the work surface that are a million times more intense than an oxyacetylene torch. Their focal points can be rapidly scanned with computer-controlled mirrors or magnetic fields to deposit energy exactly where needed. This focusing can be a tremendous advantage in, for example, heating of parts precisely at points of maximum wear, thereby eliminating the need to heat and cool the entire piece. As such, electricity can deliver packages of concentrated, precisely controlled energy and information efficiently to virtually any point. It offers society greater form value than other forms of energy since it is such a high-quality energy form. Form value affords flexibility, which in turn allows technical innovation and enormous potential for economic efficiency and growth.

A few of the specific merits of the electrotechnologies selected for analysis relative to their fossil-fueled counterparts include the following:

- **Electric Boilers:** Smaller footprint; quicker response to load changes.

- **Electric Drives:** Lower maintenance and operating cost; reduced cooling water use; improved process control.

- **Heat Pumps:** Reduced waste heat; lowered effluent temperature; improved process control; improved product quality.

- **Electrotechnologies for Process Heating:** Reduced operating and maintenance costs; improved process control; improved product quality.

CONCLUSION

One of the most important actions which can be undertaken to meet the energy needs of consumers is to make certain that end uses are as efficient as possible.

References

Healthy Buildings, Healthy People: A Vision for the 21st Century, EPA-402-K-01-003, EPA, Washington, DC: October 2001.

Lee Smith, "History Lesson: Ductless Has Come a Long Way," *ACHR News*, April 30, 2007.

Energy Market Profiles — Volume 1: 1999 Commercial Buildings, Equipment, and Energy Use and Volume 2: 1999 Residential Buildings, Appliances, and Energy Use, EPRI, Palo Alto, CA, 2001.

Clark W. Gellings, *Saving Energy with Electricity*, Discussion Paper, EPRI, Palo Alto, CA: 2008.

Phase I Reports: Potential for End Use Technologies to Improve Functionality and Meet Consumer Expectations, 2006, www.galvinpower.org.

Chapter 12

Demand-side Planning

INTRODUCTION

From the energy issues first raised in the 1970s, participants in the electric sector have learned the importance of a flexible and diverse management strategy that will help them succeed in an increasingly competitive and uncertain market. A key challenge through these decades continues to be balancing customer demand for electricity with cost, environmental concerns, and stakeholder requirements. Recognizing and planning for the integration of customer programs like energy efficiency, demand response and electrification can help mitigate the need for future generating capacity, assure efficient utilization of facilities, and augment the range of choices available to utility management in meeting the environmental challenges of the future. Table 12-1 lists the definitions used in discussing demand-side planning.

This chapter is intended to provide an overview of the key issues in that regard. This focuses on the broad issues of demand-side planning ranging from motives for considering energy efficiency, to techniques for analyzing the cost-effectiveness of alternatives, and to analyses of implementation issues. In summary, the key issues that will be developed here are:

- Demand-side planning can elucidate a broad range of alternatives for electricity demand. Thus, it warrants examination by virtually every participant in the electric sector. These options are equally applicable and desirable for investor-owned, municipal, and rural electric utilities, whether distribution, transmission or generation, as well as system operators, third-party providers or energy service entities.

- The arrangement between consumers and providers can be dynamic. There is a need to constantly assess the demand for electric-

Table 12-1. Key Definitions

Demand-Side Management – the planning and implementation of those utility activities designed to influence customer use of electricity in ways that will produce desired changes in the load shape – i.e., changes in the pattern and magnitude of a load. Demand-side management encompasses the entire range of management functions associated with directing demand-side activities, including program planning, evaluation, implementation, and monitoring. Opportunities for demand-side management can be found in all customer classes including residential, commercial, industrial, and wholesale.
Demand-Side Planning – those activities related to the design and assessment of programs directed at influencing customer use of electricity.
Demand-Side Implementation – those activities related to the installation of a demand-side program once its cost-effectiveness has been determined.
Load Shape Objectives – the changes in customer use of electricity desired in order to achieve specific goals. These desired load shape changes can be used to characterize the potential impact of alternative demand-side management programs. Although there is an infinite combination of load shape changing possibilities, six are commonly referred to illustrate the range of possibilities: peak clipping, valley filling, load shifting, strategic conservation, load growth from new efficient uses of electricity, and flexible load shape.
Load Management – utility activities designed to influence the timing and magnitude of customer use of electricity. To many people, the traditional load shape objectives of load management include peak clipping, valley filling, and load shifting.
Demand-Side Alternatives – products and services that provide the vehicle by which demand-side management programs can influence the use of electricity. Demand-side alternatives can be characterized in four ways: by the intended load shape objective of the demand-side program, by the customer end use that is affected, by the technology or equipment involved; and by the manner in which market implementation is achieved.
Demand Response – activities designed to influence consumer demand in the short term principally to reduce peak demand. Typically, includes load management direct load control, time-of-use and real-time pricing or critical peak pricing.

ity and a need to alter course as economic or operating conditions change.

- Demand-side programs and initiatives can provide customers with the opportunity they desire to better manage their total energy cost and usage and the impact they have on the environment, but in a manner that will hold mutual benefits.

- Demand-side programs and activities include many new activities, such as electrification, which can lead to increased efficiency in overall customer energy use and to improve customer productivity.

- The impacts of customer programs are highly specific, but there is a wealth of data and a very rapidly growing experience base on which to draw.

- Demand-side activities warrant the same level of attention and resources that are given to whole-market and other supply-side requirements—i.e., the years of planning and scheduling, rigorous analytical modeling, and calculations concerning reliability, operation, and maintenance.

- The extent to which market participants can successfully carry out demand-side initiatives will depend on a number of factors, none of which will be more important than the active support of top management and regulators. The essence of demand-side programs lies in how providers can relate to their customers and to the regulatory community. They must be permitted to treat demand-side activities as assets in much the same way transmission, distribution or generation assets are treated.

What is Demand-side Planning?

Demand-side planning is the planning of those activities designed to influence customer use of electricity in ways that will produce desired changes in the utility's load shape—i.e., changes in the time pattern and magnitude of a utility's load or the adoption of new electric technologies which displace fossil energy uses. Programs falling under this umbrella include demand response (such as time of use rates, load management, and direct-load control), new efficient uses, energy efficiency and distributed generation. Demand-side activities have evolved and expanded over the last few years as seen in Figure 12-1.

Demand-side activities involve a deliberate intervention by the market participant through the establishment of an infrastructure and programs so as to alter the overall pattern and/or demand for electricity. Under this definition, customer purchases of energy-efficient refrigerators would not be classified as part of a utility program. Rather, they would be considered part of naturally occurring responses of consumers to prices, and the availability of improved devices. On the other hand, a program that encourages customers to install energy-efficient refrigerators, through either incentives or advertising, meets the definition of demand-side programs. While this distinction be-

tween "naturally" occurring
and deliberately induced
changes in energy consumption and load shape is at
time difficult to make, it is
nevertheless important.

Note also that demand-
side alternatives extend beyond demand response,
energy efficiency and load
management to include programs designed specifically
to incorporate new efficient
uses of electricity which
could add load overall or in
peak and off-peak periods.
Thus, demand-side alternatives warrant consideration
regardless of the regulatory
arrangements and business
models involved.

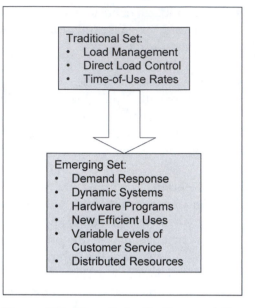

Figure 12-1. Growth in the Range of Demand-side Activities

Why Consider the Demand Side?

Why should electric energy providers be interested in customers?
Since the early 1970s, economic, political, social, technological, and
resource supply factors have combined to change the utility industry's
operating environment and its outlook for the future. Now, if 2007,
many providers are faced with staggering environmental concerns,
capital requirements for new plants, significant fluctuations in demand and energy growth rates, declining financial performance, and
regulatory and consumer concern about rising prices. While embracing
customers is not a cure-all for these difficulties, it does provide management with a great many additional alternatives. These demand-side
alternatives are equally appropriate for consideration by all electric
sector participants.

For areas with strong load growth, demand response and energy efficiency can provide an effective means to reduce the need for
wholesale capacity while minimizing the environmental footprint. For
others, deploying new uses of electricity and deliberate increases in

the market share of energy-intensive uses can improve the utility load characteristics, reduce overall environmental impacts, and optimize return. Aside from these rather obvious cases, changing the load shape that a provider must serve can reduce operating costs. Changes in the load shape can permit adjustments in short-term market purchase, generation operations, and the use of less expensive energy sources as well as in sources with lower environmental impact.

Selecting Alternatives

Finally, demand-side planning encompasses planning, evaluation, implementation, and monitoring of activities selected from among a wide variety of programmatic and technical alternatives. This is complicated by the various values customers and providers have. This wide range of alternatives mandates that providers seriously consider the demand-side by including it as a part of their overall planning process. Due to the large number of alternatives, however, assessing which alternative is best suited for a given energy service provider is not a trivial task. The choice is complicated by the fact that the attractiveness of alternatives is influenced strongly by area-specific factors, such as the regulatory environment, current generating mix, expected load growth, capacity expansion plans, load factor, load shapes for average and extreme days, and reserve margins. Therefore, it is inappropriate to transfer these varying specific factors from one service area or region to another without appropriate adjustments.

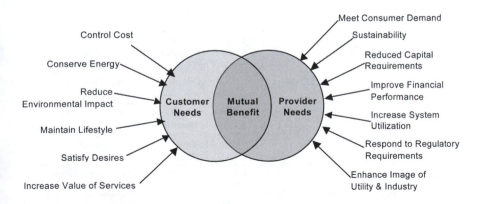

Adapted from Kyle Wilcuff, Southern Company Services, "Total Energy Resource

Figure 12-2. A Renewed Partnership

ISSUES CRITICAL TO THE DEMAND-SIDE

Figure 12-2 illustrates the needs which customers and providers perceive in demand-side programs. These infer eight critical issues that utilities considering the demand-side must resolve. The eight questions are listed in Table 12-2.

How Can Demand-side Activities Help Achieve Its Objective

Although provider needs and characteristics vary widely within the industry, every utility should examine demand-side alternatives. Embracing the demand side can offer a utility a broad range of alternatives for reducing or modifying load during a particular time of the day, during a certain season, or annually.

The Utility Planning Process

In some cases, reviews of demand response, energy efficiency, and deploy new efficient uses have emphasized the impacts of a single alternative with little discussion on how that alternative was first selected. The preferred approach in assessing the overall viability of planning is

Table 12-2. Issues in Demand-side Planning

In order to...	You need to ask...
Establish Program Objectives	How can demand-side activities help achieve my objectives?
Identify Alternatives	What type of demand-side activities should I be getting involved in?
Determine Evaluation Process	How can I select those alternatives that are most beneficial for my entity?
Forecast Load Shape Impacts	What changes in the load shape can be expected by implementing demand-side alternatives?
Develop Marketing Strategies	How can we forecast and promote customer acceptance of demand-side alternatives?
Implement Program	What is the best way to implement selected demand-side alternatives?
Monitor Results	How should my utility monitor the results of a demand-side program?
Meet Consumer Demand	Is energy efficiency the highest-priority resource for meeting demand?
Reduce Carbon Footprint	What reduction in CO_2 results from my programs?
Initiate Action	How do I get started in addressing demand-side issues as they relate to my company?

to incorporate the assessment as part of the utility's strategic planning process. Discussion here focuses on a three-level hierarchy in utility planning related to demand-side activities: establishing broad objectives, setting specific operational objectives, and determining desired load shape modifications. This is illustrated in Figure 12-3. Table 12-3 defines the load shape objectives listed.

The first level of an energy service provider's formal planning process is to establish overall organizational objectives. These strategic objectives are quite broad and generally include such examples as improving cash flow, increasing earnings, or improving customer satisfaction. Clearly, they differ depending on the position one has in the value chain (i.e., whether you are a distribution utility, transmission owner or

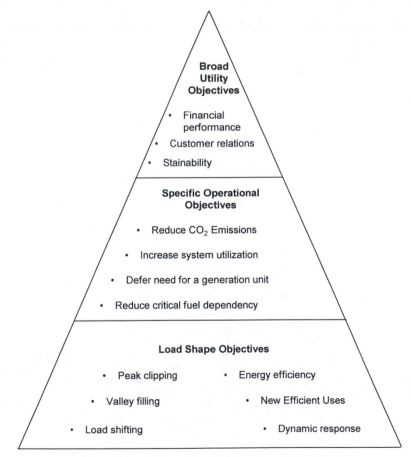

Figure 12-3. Hierarchy of Planning Objectives

Table 12-3. Load Shape Objectives*

Peak Clipping, or the reduction of the system peak loads, embodies one of the classic forms of load management. Peak clipping is generally considered as the reduction of peak load by using direct load control. Direct load control is most commonly practiced by direct control of customers' appliances. While many service providers consider this as a means to reduce peaking capacity or capacity purchases and consider control only during the most probable days of system peak, direct load control can be used to reduce operating cost and dependence on critical fuels by economic dispatch.	
Valley Filling is the second classic form of load management. Valley filling encompasses building off-peak loads. This may be particularly desirable where the long-run incremental cost is less than the average price of electricity. Adding property priced off-peak load under those circumstances decreases the average price. Valley filling can be accomplished in several ways, one of the most popular of which is new thermal energy storage (water heating and/or space heating) that displaces loads served by fossil fuels.	
Load Shifting is the last classic form of load management. This involves shifting load from on-peak to off-peak periods. Popular applications include use of storage water heating, storage space heating, coolness storage, and customer load shifts. In this case, the load shift from storage devices involves displacing what would have been conventional appliances served by electricity.	
Energy Efficiency is the load shape change that results from programs directed at end-use consumption. Not normally considered load management, the change reflects a modification of the load shape involving a reduction in sales as well as a change in the pattern of use. In employing energy efficiency, the planner must consider what conservation actions would occur naturally and then evaluate the cost-effectiveness of possible intended programs to accelerate or stimulate those actions. Examples include weatherization and appliance efficiency improvement.	
Deploying new, efficient uses is the load shape change that refers to a general increase in sales beyond the valley filling described previously. This may include electrifying existing fossil uses. These new efficient uses may include the new emerging electric technologies surrounding electric vehicles, industrial process heating, and automation. These have a potential for increasing the electric energy intensity of the U.S. industrial sector. This rise in intensity may be motivated by reduction in the use of fossil fuels and raw materials resulting in improved overall productivity and a reduced impact on the environment.	
Demand Response is a concept related to reliability, a planning constraint. Once the anticipated load shape, including demand-side activities, is forecast over the corporate planning horizon, the power supply planner studies the final optimum supply-side options. Among the many criteria used is reliability. Load shape can be flexible – if customers are presented with options as to the variations in quality of service that they are willing to allow in exchange for various incentives. The programs involved can be variations of interruptible or curtailable load; concepts of pooled, integrated energy management systems; or individual customer load control devices offering service constraints.	

operator, independent system operator, power producer, vertically integrated utility, or energy service provider with no assets). Certain institutional constraints may limit the achievement of these objectives. These
constraints represent the obvious regulatory environment that many
players face—competition, regulation, environmental considerations,
and the obligation for some to provide service of reasonable quality to
customers. While all electric sector stakeholders face such institutional
constraints, they differ between states and certainly between investor-
owned and public power utilities, as well as across system operators
and power providers.

While overall organizational objectives are important guidelines for
electric system long-range planning, there is a need for a second level of
the formal electric system planning process in which a provider's objectives are operationalized to guide management to specific actions. It is
at this operational level or tactical level that demand-side alternatives
should be examined and evaluated. For example, an examination of
capital investment requirements may show periods of high investment
needs. Postponing the need for new construction through a demand-side
program may reduce investment needs and stabilize the financial future of
the market participant. As CO_2 constraints emerge and tighten, including
demand-side alternatives, options will become only more desirable.

Specific operational objectives are established on the basis of the
conditions of the existing provider—its business model, regulatory, environmental and system configuration, case reserves, operating environment, and competition. Operational objectives that can be addressed by
demand-side alternatives include:

- Reducing the need for capacity
- Reducing the need for fossil fuels
- Reducing CO_2 emissions
- Reducing or postponing capital investment
- Controlling electricity costs
- Increasing profitability
- Providing customers with options that provide a measure of control
  over their electric bills
- Reducing risks by investing in diverse alternatives
- Increasing operating flexibility and system reliability
- Decreasing unit cost through more efficient loading of existing and
  planned generating facilities

- Satisfying regulatory constraints or rules
- Increasing sustainability
- Improving the image of the utility

Once designated, operational objectives are translated into desired load shape changes that can be used to characterize the potential impact of alternative demand-side alternatives.

The potential role that can be filled by these demand-side alternatives in a planning process looks rather ambitious. Nevertheless, the demand-side approach does provide management with a whole new set of alternatives with which to meet the needs of its customers. The concept that consumer demand is not fixed but can be altered deliberately with the provider, regulator, and customer cooperating opens a new dimension in planning and operation.

Demand Response & Energy Efficiency

The obvious question in response to the above claims is: Can demand-side activities help achieve the broad range of operational objectives listed by merely changing consumer demand? The answer is that numerous industries have found that changing the pattern of the demand for their product can be profitable. For example, telephone utilities have long offered reduced evening rates to shift demand and to encourage usage during non-business hours. Airlines offer night coach fares to build traffic during off-peak hours. Movie theaters offer reduced matinee prices to attract additional customers. All of these examples are deliberate attempts to change the demand pattern for a product to encourage efficient use of resources and thus profitability.

What Type of Demand-side Activities Should Providers Pursue?

Although customers and providers can act independently to alter the pattern of demand, the concept of demand-side planning implies a relationship that produces mutually beneficial results between providers and customers. To achieve these mutual benefits, a provider must carefully consider such factors as the manner in which the activity will affect the load shape, the methods available for obtaining customer participation, and the likely magnitudes of costs and benefits to both provider and customer prior to attempting implementation.

Because there are so many demand-side alternatives, the process of identifying potential candidates can be carried out more effectively by considering several aspects of the alternatives in an orderly fashion. Demand-side activities can be categorized in a two-level process in which the second level has three steps and illustrated in Figure 12-4.

Level I: •   Load Shape Objectives
Level II: •   End Use
           •   Technology Alternatives
           •   Market Implementation Methods

The first step in identifying demand-side alternatives is typically the selection of an appropriate load shape objective to ensure that the desired result is consistent with utility goals and constraints. For example, load forecasts may indicate that existing and planned generating

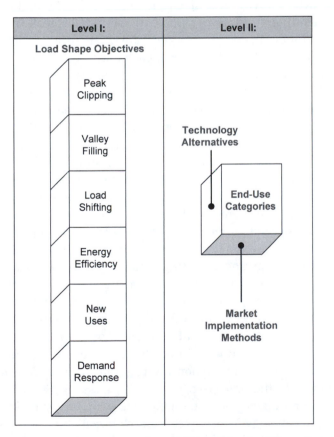

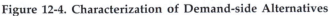

Figure 12-4. Characterization of Demand-side Alternatives

capacity will fall short of projected peak demand plus targeted reserve margins. Several supply-side alternatives may be available to meet this capacity shortfall: additional peaking capacity can be built; extra power can be purchased as needed from other generating utilities; or perhaps, a reduction in reserve margin can be tolerated. There are also a number of demand-side alternatives, including demand response, direct load control, interruptible rates, and energy storage, that can augment the number of planning alternatives available to a utility. Choosing between meeting peak versus reducing the peak becomes a balance between the costs and benefits associated with the range of available supply-side and demand-side alternatives.

Once the load shape objective has been established, it is necessary to find ways to achieve it. This is the second level in the identification process which involves three steps or dimensions. The first dimension involves identifying the appropriate end uses whose peak load and energy consumption characteristics generally match the requirements of the load shape objectives. In general, each end use (e.g., space heating, lighting) exhibits typical and predictable load patterns. The extent to which load pattern modification can be accommodated by a given end use is one factor used to select an end use for demand-side planning.

There are nine major end uses which have the greatest potential. They are space heating, space cooling, water heating, lighting, refrigeration, cooking, laundry, swimming pools, and miscellaneous other uses. Each of these end uses provides a different set of opportunities to meet some or all of the load shape modification objectives that have been discussed. Some of the end uses can successfully serve as the focus of programs to meet any of the load shape objectives, while others can realistically be useful for meeting only one or two of these objectives. In general, space heating, space cooling, and water heating are the residential end uses with the greatest potential applicability for achieving load shape objectives. These end uses tend to be among the most energy intensive and among the most adaptable in terms of having their usage pattern altered. However, some energy service providers and distribution utilities have achieved significant load shape modifications by implementing programs based on or including combinations of other end uses.

The second dimension of demand-side planning involves choosing appropriate technology alternatives for each target end use. This process should consider the suitability of the technology for satisfying the load shape objective. Even though a technology is suitable for a given end use, it may not produce the desired results. For example, although heat pumps are appropriate for reducing domestic water heating electric consumption, they are not appropriate for load shifting. In this case, an option such as direct load control would be a better choice.

Residential demand-side technologies can be grouped into four general categories as described in Table 12-4:

* Building envelope alternatives
* Efficient equipment and appliances
* Thermal storage equipment
* Demand Response (energy and demand control options, including dynamic response

These four main categories cover most of the currently available, or soon to be available, customer options. Many of the individual options can be considered as components of an overall program and thereby offer a very broad range of possibilities for successful residential demand-side program synthesis and implementation. These options are described in greater detail in Table 12-4.

The third dimension of demand-side alternatives deals with various methods for encouraging the customer to participate in the program. Market implementation methods vary for different technologies. Frequently, two or more customer adoption strategies are used simultaneously to promote a given program. The different types of customer adoption techniques represent varying levels of utility involvement. Direct incentive programs, for example, represent a high degree of utility support in promoting demand-side programs. Customer awareness strategies can require less utility involvement.

Taken in sequence, the four steps of activities described provide an orderly method for characterizing demand-side management alternatives. The basic steps are:

* Establish the load shape objective to be met

* Determine which end uses can be appropriately modified to meet the load shape objectives

* Select technology options that can produce the desired end use load shape changes

* Identify an appropriate market implementation plan program

Table 12-4. Residential Demand-side Technology Alternatives

Efficient Equipment & Appliance Alternatives		Thermal Storage Equipment
Heat Pumps	**Dual Fuel Heating System**	**Heat Storage**
– Central Air Source Heat Pump	– Dual Fuel Heating System	– Central Ceramic Heat Storage
– Ground-Water Source Heat Pump	– Add-On Heat Pump	– Room Ceramic Heat Storage
– Ground-Coupled Heat Pump	– Active Solar Space Heating	– Slab Heating
– Multi-Zone Heat Pump	– Task Heating	**Cool Storage**
– Room Heat Pump	– Zoned Resistance	– Residential Ice Storage Air Conditioning
High-Efficiency Appliances	**Water Heating Equipment**	**Demand Response**
– High-EER Air Conditioner	– Heat Pump Water Heater	– Receiver Switches
– Energy-Efficient Cooking Appliances	– Heat Recovery Water Heater	– Water Heater Cycling Control
– Energy-Efficient Washers & Dishwashers	– Solar Water Heating	– Air Conditioner Cycling Control
– Energy-Efficient Refrigerators & Freezers		– Local Utility or Customer Control
– Efficient Lighting Fixtures & Lamps		– Variable-Service-Level Devices
Building Envelope Alternatives		– Timers
Thermal Treatment	**Infiltration & Indoor Air Quality**	– Appliance Interlocks
– Insulation (Ceilings, Walls, Floors)	– Infiltration & Indoor Air Quality Control	– Programmable Controllers
– Storm & Thermopane Windows, Storm Doors		– Temperature-Activated Time Switches
– Window Treatments (Shades, Solar Screens)	**Passive Solar Design & Day Lighting**	– Load Management Thermostats
– Duct & Pipe Insulation	– Passive Solar Design	– Swimming Pool Pump Control
– Water Heater Blanket	– Day lighting	– Demand Response Alternatives
		– Dynamic Systems
		– Time-of-Use Pricing

How Do I Select Those Alternatives That Are Most Beneficial?

Selection of the most appropriate demand-side alternatives is perhaps the most crucial question a service provider faces. The question is difficult since the number of demand-side alternatives from which to select is so large. In addition, because the relative attractiveness of alternatives depends upon specific characteristics, such as regulatory, environment, load shape summer and winter peaks, generation mix, customer mix, and load growth, transfer of results from one service area to another may not be appropriate. In other words, what is attractive to one utility may not be attractive to another.

Completing detailed evaluations of demand-side programs can be complex and may even appear overwhelming. These evaluations typically require a great deal of data and a computer model for processing. However, a detailed analysis of demand-side alternatives is not the starting point in the selection process.

Reference

Demand-side Management Planning, C.W. Gellings and J.H. Chamberlin, PennWell Publishing Co., 1993.

Chapter 13

Demand-side Evaluation

LEVELS OF ANALYSIS

Because there are so many different demand-side alternatives available, they should be analyzed through a hierarchy of evaluation levels, starting with an intuitive selection, continuing with an aggregate analysis, and ending with a detailed and comprehensive evaluation. To a large extent, the appropriate level of analysis depends upon the importance of the decision that will be influenced by the analysis. In the hierarchy, quick and less demanding analysis is used to identify the most attractive candidates for more extensive analysis. However, the analyst must ensure that the potential value of additional, more detailed analysis is not outweighed by the cost of completing the detailed analysis. This is illustrated in Figure 13-1.

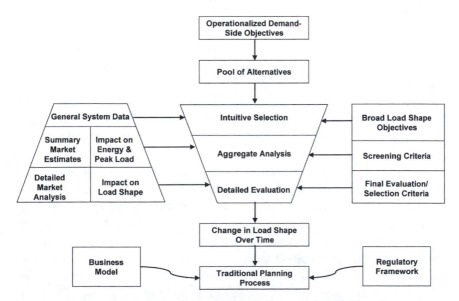

Figure 13-1. Levels of Evaluation in Demand-side Planning

259

Although every service provider does not need to go through all three levels of analysis or even the final level in selecting an alternative, the starting point should be an intuitive selection of those alternatives that, without extensive analysis, seem to satisfy the provider's and the utility's needs. For example, a provider concerned about its summer peak and low load factor could be investigating heat pumps to build winter loads or demand response programs to reduce the summer peak growth. Options such as weatherization or thermal storage may be of lesser interest.

The first level, intuitive selection, is based upon a thorough understanding of the conditions within the service area, of the wholesale electricity market or native generation portfolio and planned expansion, and of the operating characteristics of the demand-side alternatives. Note that the intuitive selection process does not identify those alternatives that are in some sense "best" for the service area. Rather, the process identifies a number of alternatives that are, at least initially, appropriate to achieve stated goals.

The next level in identifying alternatives is a more quantitative analysis that examines costs and benefits to all parties affected by implementation of a specific program. Interested parties include the service provider, third-party providers, the participants in the program, other customers, and often society at large. To calculate the costs and benefits requires quantitative information on the impact of the alternative on peak, on the pattern of demand, and total energy sales, the expected participation in the program, the costs of implementation, and wholesale market or generating system data (such as costs for existing units). For this level of analysis, all expected costs and benefits to the provider, the program participants, and the public at large are projected for the entire life of the program. Comparison of the benefit/cost ratios will then yield preliminary ranking of programs. More detailed rankings can be made of combined programs.

The final step in the selection of the most appropriate demand-side alternative is a detailed analysis of the most cost-effective alternatives. In a typical detailed analysis, the performance of the electric system from both an operational and a financial viewpoint is stimulated over time with and without the selected demand-side alternative. This analysis estimates changes in wholesale market prices or in the generating system and its operation that will result from the altered load shape produced by the selected demand-side alternatives. It builds heavily on the region's existing analysis tools and cooperate models.

GENERAL INFORMATION REQUIREMENTS

Such analysis requires a great deal of information on the existing and anticipated wholesale market, any native generation, and on the demand-side program. While the information describing the current and anticipated wholesale market or the planned generating system and its operation is generally available from capacity expansion and production costing analyses, obtaining the information for demand-side programs is often a challenge. Specifically, information on the load shape of the end use and changes in that load shape resulting from the implementation of the selected demand-side alternative is required. In addition, it is necessary to characterize the end use in the service area. In the case of space heating, for example, this implies accumulating information on the number of electric space heating customers, the annual heating requirements, the type of equipment (e.g., electric furnace, resistance baseboard heat, or heat pump) and likely future changes in heating requirements. Finally, projections of customer acceptance and response to the program, taking into account customers' costs and savings, must be developed for the planning horizon. Since much of the information needed is related to the customer, marketing departments can plan a major role in gathering it.

Implicit in the selection process is a definite strategy to reduce the information requirements to manageable levels consistent with the trade-off between the data collection/analysis expense and the resulting level of accuracy in the evaluation. This strategy focuses on quickly and efficiently reducing the number of alternatives appropriate for both a given service provider and geographic area. Although the detailed analysis is the most comprehensive and in some sense the most "accurate" assessment, the amount of information prohibits its use on all potential alternatives.

Thus, the selection process starts with informed personnel selecting those alternatives that seem most appropriate based upon insight into their service area. Since demand-side alternatives focus on the customer, or more correctly on a customer's end uses, customer-related background information is essential in this first step. The next step, the preliminary cost/benefit analysis, requires more quantitative information but for only those alternatives that have some promise for the service area. Finally, the comprehensive analysis is applied only to those alternatives that have the highest benefit-to-cost ratio.

The remainder of this discussion focuses on a number of important concerns in the evaluation process. Some of these are listed in Table 13-1.

Table 13-1. Concerns in Evaluating Demand-side Alternatives

• **Evaluate alternatives in conjunction with supply-side options or wholesale market prices**
• Be careful in transferring results from one service area to another
• Be aware that detailed analyses are data intensive
• Expand use of cost/benefit analysis method to evaluate demand-side alternatives
• Consider factors other than monetary cost

SYSTEM CONTEXT

Demand-side alternatives must be evaluated in the context of the supply-side alternatives—the supply system. Just as in the analysis of supply-side alternatives, the attractiveness of alternatives depends upon the existing and planned supply system and on the characteristics of the alternatives themselves. Demand-side alternatives alter the load shape and thus affect the operating efficiency and future capacity additions. Translating these effects into specific cost savings depends upon the characteristics of the supply system.

TRANSFERABILITY

Caution must be exercised in transferring results of a selection process from one service area to another. While the analysis techniques, and in some cases data, can be transferred, in most cases, the results cannot. There are a number of reasons for this. First, cost savings resulting from the implementation of demand-side alternatives depend upon the provider's load shape and the wholesale market or generating system it has been designed to serve. Most often, this is a unique combination. Second, customer end-use characteristics often differ between service areas. For example, electric space heating and electric water heating saturation levels are often quite different, even for providers in the same geographic area. Thus, while the impact of an alternative may be the same on a per unit basis, the total impact resulting from the same saturation will be quite different. Customer characteristics and climatic conditions also influence the impacts on a per unit or per customer basis. Energy savings resulting from heating, cooling or weatherization programs change, depending upon dwelling type (e.g., single-family vs. multi-family), dwelling size, and climatic conditions measured (e.g., by heating degree days).

DATA REQUIREMENTS

Detailed analyses of demand-side alternatives are data intensive, requiring information in four major categories:

- Service area-specific customer and end-use characteristics (type of equipment in use, stock estimates of this equipment, patterns of usage).
- Operating/technical characteristics of the alternatives.
- Characteristics of the supply system (operating costs, reliability, initial cost).
- Customer acceptance of alternatives.

It is often said that demand-side planning data tend to be "softer" and much more customer oriented than supply-side data. Because supply-side data tend to be much more hardware and engineering oriented, they give the impression of greater reliability. However, since supply-side planning is based on projected future energy requirements (i.e., customer demands), it actually involves the same uncertainties. Moreover, critical variables in supply-side planning are often known with no more accuracy than demand-side variables.

COST/BENEFIT ANALYSIS

The cost/benefit evaluation approach is the preferred approach to assess demand-side alternatives. Although analysts typically minimize future revenue requirements in evaluating supply-side alternatives, this measure is inappropriate for evaluating some demand-side alternatives. For example, a program designed to build off-peak use of electricity, such as an add-on heat pump program, will increase revenue requirements. However, increased utilization of existing capacity will lower the unit cost of power—clearly a benefit to both the customers and the utility. Regardless of whether the analysis uses required revenues or unit costs as the measure of program impact, the costs and benefits of serving the load shape, with or without the demand-side alternative is place, must be estimated and compared over the planning horizon. In this comparison, it is important to perform the analysis on an equivalent unit basis. Thus, two systems are assumed to be equivalent if they serve the load at the same level of reliability. For utilities with excess reserves attempting to build load, it is appropriate to perform the analysis on the desired reserve level. In cases where the reserve levels are not equal, some estimates for the cost of outages must be included in the analysis.

NON-MONETARY BENEFITS & COSTS

Factors other than easily quantified monetary benefits and costs are often important in the selection of demand-side alternatives. Although the discussion has dwelt upon savings in monetary costs up to this point, there are other important factors influencing the selection of these alternatives. Among these are:

- Potential use of CO_2 trading
- Cash flow
- Magnitude of start-up
- Public relations reaction
- Compatibility of alternatives
- Uncertainty/risk associated with success of program
- Availability of competent installation/products
- Customer reaction/participation
- Real or perceived customer health/safety issues
- Ease/convenience of service installation
- Conformance to building codes and standards
- Regulatory/institutional limitations and opportunities

Because such variables are difficult to quantify, they are typically not included in a formal quantitative analysis. Instead, they are incorporated qualitatively in the summary assessment of the proposed program. However, there is increasing interest in explicitly quantifying CO_2 savings from demand-side programs.

What Changes in the Load Shape Can Be Expected By Implementing Demand-side Alternatives?

As outlined in the introduction, demand-side planning focuses on deliberately changing the load shape so as to optimize the entire power system from generation to delivery to end use. This focus may give the impression that they only load shape changes that occur are those induced by demand-side programs and activities. Certainly this is not the case. System load shape changes can occur naturally due to fluctuations in customer mix, the entry of new industries into the marketplace, the introduction of new processes, and the growth of end-use stock in the residential and commercial sectors. Thus, to examine the impact of demand-side alternatives, it is important to differentiate naturally occurring changes in the load shape and those changes resulting from demand-side alternatives.

Customers purchase electricity to satisfy a need for energy, not to meet utility demand growth. The residential load shapes are influenced by the customer's decision to purchase an appliance, as well as the resulting level of use of that appliance. Similarly, in the commercial and industrial sectors, installation of equipment and its utilization affect the load shape. Projections of the purchase of appliances and the behavior of customers combine to produce a forecast of the residential load shape. Numerous factors influence both the selection and utilization decision, including:

- Price of electricity and competing fuels
- Demographics (income, age, and education)
- Appliance characteristics (saturation, usage, cost and age)
- Behavioral factors
- Utility marketing/program availability
- Mandated standards
- Government programs

The action taken by consumers once an appliance or device has been purchased and installed, combined with the design characteristics of the device, results in the load shape change or the customer response as illustrated in Figure 13-2.

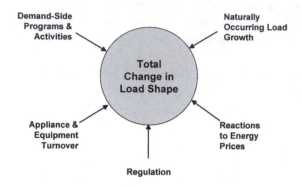

Figure 13-2. Factors Influencing System Load Shape

The same basic mechanism used to describe the load shape and its change over time can also be used to estimate changes in that load shape due to demand-side programs and activities. Changes in an individual customer's load shape result from two different factors:

- Changes in customer utilization of existing appliances or equipment (e.g., installation of clock thermostat setback in response to an advertising campaign or conservation).

- Changes in the operating characteristics or technology for a given end use (e.g., substitution of a heat pump for an electric furnace).

Obviously, these factors are not mutually exclusive and often interact. System load shape changes are the cumulative response of individual customer load shape changes plus the load contributed by new customers.

While both of the above factors produce changes in the load shape, it is important to differentiate the effects of each. In the case of behaviorally induced changes, the emphasis is on how the customer responds and how permanent that response is. Since in many cases there is no customer investment, the changes may only be temporary, with customers eventually reverting back to their original behavior patterns.

The technology-induced changes in the load shape depend upon the relative usage level of the device, as well as its improved design characteristics. As such, the changes in the load shape tend to be "permanent" and more predictable. Returning to the heat pump example, the load shape may still change as the customer becomes more energy conscious or responds to a demand response program. However, these changes are limited when compared to the change resulting from the improved operating characteristics of an advanced replacement heap pump. The critical question remains—what is the level of customer acceptance of the device? Acceptance of demand-side alternatives is discussed in more detail later.

Load shape changes that can be expected from the implementation of demand-side alternatives vary. This should not be surprising since factors unique to the service area influence changes in the system load shape.

The remainder of the discussion in this section focuses on two critical issues related to load shape impacts: program interaction and dynamic systems.

PROGRAM INTERACTION

Alternatives tend to interact making the estimation of changes in the load shape difficult. Demand-side alternatives are often promoted and implemented as an integrated program. Promotion of heat pumps not only affects heating requirements, but also cooling loads. Moreover, if that heat pump program is coupled to a weatherization program, energy requirements are lowered for those customers participating in both programs. Similarly, there may be changes resulting from cycling air conditioners that can be influenced by the use of a "smart" thermostat. The kind of thermostat "knows" the future power interruption and automatically lowers the indoor temperature in anticipation of expected reduced air conditioner operation. Interactions of alternatives must be carefully examined and their implementations analyzed.

DYNAMIC SYSTEMS

Load shape impacts of demand-side alternatives may change over the planning horizon. Just as the system load shape is dynamic and is expected to change over time, so are the load impacts of many demand-side alternatives. Failure to recognize and account for this can lead to serious future supply problems. For example, the amount of control that can be exercised over the cooling load using direct load control without a dynamic system will change over the course of the planning horizon. For many utilities, the stock of central air conditioners is still growing due to increasing saturation levels and/or customer growth. Therefore, assuming the same level of participation in the program, the total amount of control tends to increase. On the other hand, customers are replacing old and inefficient units with much more energy efficient ones. Assuming the same amount of control, these efficiency improvements would tend to decrease the impact of the load control interruptions. The net effect of these opposing trends depends on the specific conditions present in the service area. The important message, however, is that the load shape is dynamic and changes over the planning horizon. Table 13-2 gives examples of load shape changes resulting from select demand-side alternatives.

How Can Adoption of Demand-side Alternatives Be Forecasted and Promoted?

For many years, energy services departments successfully stimulated company growth by directing numerous advertising and sales efforts such as the "Live Better Electrically" Program conducted in cooperation with appliance manufacturers. But, as electricity costs accelerated and national targets for utility use of petroleum and natural gas were set, supply-side issues took on a greater importance. During the 1970s and 80s, many of these marketing efforts were dismantled and the marketing staff dispersed to other departments. It has only recently (albeit modestly) been recognized nationally that marketing can be used to shape load as well as to stimulate it. Some customer-related departments can make valuable contributions to developing demand-side solutions to load problems and to taking advantage of load opportunities. Thus, one of the first and most fundamental issues to be addressed in the development of marketing programs is to locate and pull together the expertise and tools necessary to identify, evaluate, and implement successful marketing programs.

It is an old adage that "nothing happens until someone sells something." In the demand-side planning context, this means that even if the best analysis techniques have been applied, data collected, and technol-

Table 13-2. Typical Load Shape Changes Resulting From Select Demand-side Alternatives

Energy & Demand Control Equipment	**Direct Load Control of Residential Water Heaters** – Reduction in coincident demand ranging from 0.3 kW to 1.6 kW per control per unit with a clustering of reduction around 0.5-1.0 kW.	**Load Shifting**	
	Direct Load Control of Residential Central Air Conditioners – Reduction in coincident demand ranging from 0.6 kW to 2.0 kW per unit – averaging around 1.0 kW.	**Peak Clipping**	
Thermal Storage Equipment	**Central Ceramic Heat Storage** – Can completely eliminate peak period heating demands (4 kW to 6.5 kW in test homes) and result in increases of off-peak demand ranging from 200% to 330% of conventional equipment demands.	**Valley Filling**	
Efficient Equipment & Appliance Options	**Add-on Heat Pump** – In cold climate can provide roughly two-thirds of customers' heating requirements. Sales are basically off-peak.	**New Efficient Uses**	
	Dual-Fuel Heating Systems – Can produce both valley filling with new customers and flexible load shape with existing customers. Typical reported demand impacts range from 7-10 kW. All new energy consumption is completely off-peak.	**Demand Response**	
Building Envelope Alternatives	**Incorporating Passive Solar Design in Residential Buildings** – Energy savings can range from 25 to 40% of annual space conditioning consumption compared to conventional designs. Variations are due to designs and climate.	**Energy Efficiency**	

ogy developed, the success of demand-side activities often hinges on the ability to persuade customers to actively participate in the program.

It is important for utilities to understand how these decisions are reached. Customers do not purchase energy for the sake of consuming it, per se, but instead are interested in the service it provides. This service brings warmth, cooling, artificial illumination, motive power, or other conveniences.

As discussed in the previous section, potential customer interest in a demand-side program or activity may be based on a number of factors including:

- Price of electricity and competing fuels

- Demographics (income, age, and education)

- Appliance characteristics (saturation, usage, cost and age)
- Behavioral factors
- Utility marketing/program availability
- Mandated standards

Providers must also be sensitive to the fact that the stage a consumer has reached in adopting a new product or service has a bearing on the type of marketing that should take place.

The goal of the programs and their associated marketing effort is often to increase the engagement in the demand-side program or activity. To make these successful, it is important to know the current market penetration of certain end-use devices. If market penetration is deemed too low, a number of options are worth considering. The provider may:

- Attempt to expand the number of people comprising the target market.

- Enlarge the number of devices covered in program to approach the total market potential.

- Increase or modify advertising/promotional efforts to encourage greater response from the existing market.

Caution should be used, however, in setting goals near the upper limit of market potential due to the possible diminishing returns of marketing expenditures over time. The concept of "optimal market share" can be explored if the utility can quantify the benefits associated with alternative market share levels.

ESTIMATING FUTURE MARKET DEMAND &
CUSTOMER PARTICIPATION RATES

Projections of future market demand depend on a large number of factors, including:

- Availability, frequency, timing, and use of programs and promotions

- State of the economy
- Seasonality
- Value of program incentives

Market demand can be projected by several methods, including:

- Buyer intention surveys
- Middleman estimates
- Market tests
- Time series analysis
- Statistical demand analysis

Buyer intention surveys solicit buying intentions for an upcoming period from a sample of target consumers. However, buyers having less clear intentions and those unwilling to report their intentions affect the reliability of results. Simple "yes—no" questions can be asked or a "full purchase probability scale," which is similar to a Likert Scale in refining customer responses, may be used.

Middleman estimates consist of information supplied by dealers, sales personnel, and field representatives regarding equipment sold and devices placed into the field.

Market tests/pilot programs are particularly appealing where a new product or service in planned and other forms of measurement are not appropriate. Where there is no prior experience with a program, pilot programs are developed and implemented to either a small part of a provider's service territory or informally offered to customers as part of another program in order to measure and evaluate customer responses.

Time series analyses rely on historical buyer data to estimate future consumptive decisions. A number of components of historical information are often analyzed, including trends, cycles, and seasons. Some use this technique for operational programs that have stable customer response rates.

Statistical demand analyses include additional variables that may either co-vary or cause a change in the effect of product and service requests. Price, income, population, construction starts, and household formation are the types of variables used. Sales, audits, installations, etc., are viewed as dependent variables and price, households, etc., are the independent variables.

CONSUMER & MARKET RESEARCH*

Secondary and/or primary market data are collected and analyzed as inputs to the market analysis methodologies. Regardless of the type of research data selected, the estimator of market penetration must consider these factors:

•   The accuracy of the data and its relevance to the estimating task at hand.

•   The cost (time and money) to collect and analyze the data.

•   The accuracy of the resulting estimate.

Primary data are collected to answer specific marketing or estimating needs. Primary data collection is usually categorized in two ways: qualitative and quantitative research. Frequently, qualitative research provides the foundation for quantitative research by clarifying relevant concepts and developing research hypotheses.

The two most common forms of qualitative research are in-depth interviewed and focus group discussions. Both techniques are used to explore issues, identify motivations and behavior, and develop an understanding of the consumer.

Quantitative research can be further subdivided into observational techniques, experimentation, and surveys. Observational techniques include a variety of unobtrusive measures, for example, utility metering of electrical use. Certain archival records generated by market participants and other end-use technology-related industries might also be classified as observational. Experimentation includes test marketing, concept testing, and a variety of other relatively powerful techniques designed to establish causality. The concept of control groups is central to experimentation. Surveys can provide data on socioeconomic characteristics, attitudes, behavior, and buying intentions. The information can be used to estimate market potential, to forecast sales, or to analyze consumer preferences.

---

*Adapted from Resource Planning Associates, Inc., *Methods for Analyzing the Market Penetration of End-Use Technologies: A Guide for Utility Planners*, published by the Electric Power Research Institute, October 1982,m EPRI EA-2702 (RP2045-2). EPRI has also recently funded a project on "Identifying Consumer Research Techniques for Electric Utilities" RP1537)

Secondary sources include data collected for some purpose other than estimating. For example, the U.S. Department of Commerce, Bureau of Census, collects data on population and housing that can be used to estimate the penetration of central air conditioning in residential buildings. A further delineation of secondary data research concerns the source—internal or external. Internal secondary data include proprietary price and sales records. External data are usually defined as published information.

The use of secondary data in estimating market penetration has advantages and disadvantages. These data usually do not answer the precise estimating needs, and there may be problems associated with the way the data were originally collected or are presented. On the other hand, secondary data are easily available, usually less expensive, and can serve to define the phenomenon.

CUSTOMER ADOPTION TECHNIQUES

Executives have a number of market implementation methods from which to choose. Most of these program options fall into one of the following categories:

- Alternative pricing (rate structures)

- Direct incentives

- Customer education

- Direct customer contact

- Trade ally cooperation

- Advertising and promotion

Table 13-3 presents a number of specific program options within these categories.

Many of these alternatives have been used successfully in the past. The selection of the incentives typically depends on a provider's prior experience with similar programs, the receptivity of state regulatory authorities, and the organizational philosophy of the utility.

Table 13-3. Examples of Customer Adoption Techniques

Customer Adoption Technique	Objective	Specific Alternatives
Customer Education	Increase customer awareness of programs	Bill inserts, brochures, information packets, displays, clearinghouses, direct mailings.
Direct Customer Contact	Through face-to-face communication en-courage greater custo-mer response to programs.	On-site energy service audits, workshops/energy clinics, store fronts/vendor sales and service.
Trade Ally Cooperation (i.e., architects, engi-neers, appliance dealers, heating/ cooling contractors	Increase capability in marketing and imple-menting programs.	Cooperative advertising and marketing, training, certification, selected product sales/service.
Advertising & Promotion	Increase public aware-ness of new programs, influence and control customer response.	Mass media (radio, TV, and newspaper) point-of-purchase advertising.
Alternative Pricing	Provide customers with pricing signals that are reflective of real eco-nomic costs and encourage a desired market response.	• Demand rates – rates based on the maximum kilowatt usage of a customer; the rates thus provide an incentive for customers to improve their load factor. • Time-of-use rates – rates where higher costs are incurred by the customer for using during a utility's peak period and lower costs during off-peak periods. • Off-peak rates – rates priced to reflect lower off-peak costs which offer customer service for specific end uses such as storage heating or storage water heating. • Seasonal rates – rates where the season in which the utility reaches its peak has a higher flat rate than other seasons. • Inverted rates – rates where consumers pay more for each unit of electric consumed in later tail blocks. The first block may or may not consist of a lifeline rate. • Variable levels of service rates where customers subscribe to a minimum electric service consistent with their needs – e.g., interruptible rates and other demand-response programs. • Promotional rates – rates designed to attract targeted groups of customers to a service area for the purpose of encouraging economic development. • Conservation rates – reduced rates based on a customer's dwelling meeting minimum energy efficiency standards, including mechanical systems.
Direct Incentives	Reduce up front purchase price and risk of hardware investments to the customer and increase short-term market penetration.	• Low/no interest loans – loans issued to customers below the current lending rate with the length of time for repayment varying. • Case grants/rebates/buy back – money paid to customers based on some criteria, usually the efficiency of the device, energy/demand saved, and difference in utility average and marginal costs. • Subsidized installation/modification –arranged demand-side options installed for a reduced fee or free of charge.

What is the Best Way to Implement Selected Demand-side Programs?*

With a few exceptions, many of the current programs being implemented are either "pilot" programs or larger efforts that are at a preliminary stage. Only a limited amount of information has been compiled on major program implementation experiences.

In this overview, program implementation refers to carrying out demand-side programs after their cost-effectiveness has been determined during the demand-side program planning and evaluation phase. Program implementation involves the many detailed day-to-day decisions that must be made to realize the goals of demand-side management programs.

PROGRAM IMPLEMENTATION ISSUES

Implementing demand-side programs involves three major considerations:

- Program Planning
- Program management
- Program logistics

Program Planning

As with any sophisticated program, a demand-side program should begin with an implementation plan. The plan includes a set of carefully defined, measurable, and obtainable goals. A program logic chart can be used to identify the program implementation process from the point of

---

*Much of the material presented in this section is adapted from:

Linda Finley, "Load Management Implementation Issues," December 1982, presentation made at the EPRI Seminar on "Planning and Assessment of Load Management."

Synergic Resources Corporation, *Electric utility Sponsored Conservation Programs: An Assessment of Implementation Mechanisms* (forthcoming. Electric Power Research Institute, RP2050-11.

Energy Management Associates, Inc., *Issues in Implementing a Load Management Program for Direct Load Control*, March 1983. Published by the Electric Power Research Institute, Report No. EPRI EA-2904 (RP2050-8).

Energy Utilization Systems, Inc., *1982 Survey of Utility Load Management, Conservation, and Solar End-Use Projects*, November 1982. Published by Electric Power Research Institute, Report No. EM-2649 (RP1940-1).

customer response to program completion. For example, in a direct load control program, decision points, such as meeting customer eligibility requirements, completing credit applications, device installation, and post-inspection can be defined. Actual program implementation can be checked against the plan and major variances reviewed as the occur.

The careful planning that characterizes other operations should carry over to the implementation of demand-side programs. The programs are expensive and prudent planning will help assure program efficiency and effectiveness. The variety of activities and functional groups involved in implementing demand-side programs further accentuates the need for proper planning.

Program Management

The implementation process involves many different functional groups or departments within the service provider. Careful management is required to ensure efficient implementation. Managing the needed widespread activities requires a complete understanding and consensus of program objectives and clear lines of functional authority and accountability.

Ongoing program management is also extremely important. The need for cost accounting, monitoring employee productivity and quality assurance should be addressed; the use of direct incentives necessitate close monitoring of program costs. For whatever reason, if periodic status reports are required, the requisite input data and the reporting of key performance indicators must be carefully included.

Program Logistics

Program support includes staffing, equipment, facilities, and training requirements. A program implementation manual is a useful tool to provide program personnel with necessary policy and procedure guidelines. The sample list of functional responsibilities in an implementation program (see Table 13-4) gives an indication of the activities that may be included in such a manual.

A customer adoption plan that coordinates the use of mass media and other advertising and promotional activities (such as bill inserts and direct mail) should be carefully integrated into the implementation program. It is always important to establish good rapport with customers, even more so now, because the industry is entering a new era—the era of providing energy services. This requires closer interaction with

Table 13-4. Sample Functional Responsibilities in Marketing Program Implementation

Function Area	Key Responsibilities
CEO/Senior Management	• Decision to implement demand-side planning • Approve program budgets and large-scale purchases and rate designs
Demand-Side Committee	• Mobilize and manage staff team • Address customer needs
Rates	• Design rate incentives for pilot and full-scale programs • Obtain regulatory approvals
Finance Planning and Billing	• Develop program budgets • Modify billing software
Distribution Planning	• Assess program impacts on distribution system
Transmission/Bulk Power Planning	• Assess program impacts on transmission system
Generation/Resource Planning	• Assess program impacts on generation system
System Operation	• Develop specs for any energy management system (EMS) that may be needed • Operate EMS
Substation/Distribution Engineering	• Develop specs for pilot and large-scale distribution supervisory system that may be needed • Operate the distribution supervisory system
Customer and Public Relations	• Design PR programs for recruiting customers for pilot and large-scale programs • Provide interface between customer and utility
Load Research/Economics and Statistics/Load Forecasting	• Monitor load research data and program performance
Purchasing	• Negotiate contracts and monitor delivery of equipment for pilot and large-scale programs
Installation/Construction/Engineering	• Train staff on installations, maintenance, and metering of equipment • Schedule customer installation • Schedule repair and maintenance
Regulatory/Legal Affairs	• Analyze potential program liability • Negotiate contractual matters

customers, and good rapport is necessary regardless of the demand-side program. Effective marketing and public education campaigns, as well as quick response to customer concerns, will help achieve this goal. Customer concerns should be addressed at all levels of program design and implementation.

Many demand-side programs and activities include installing a specific piece of equipment or hardware that will alter customer energy use to benefit both the customer and the utility. In this report, such equipment has been termed "technology alternatives" to contrast it with the other three dimensions or aspects of demand-side alternatives—the intended program objective, the affected end use, and the selected customer adoption technique or marketing strategy.

Some of these technology alternatives are installed, used, or marketed as part of the demand-side program. The special issues related to such demand-side programs involving utility or third-party ownership and installed equipment include selecting the proper equipment or hardware, establishing an appropriate customer adoption program, developing quality assurance programs, and developing an installation and maintenance schedule.

Selecting the Proper Equipment or Hardware

Stakeholders need to evaluate a variety of conflicting factors if they are specifying the functional requirements for equipment or hardware. The specifications for any new equipment must be coordinated with existing customer and utility hardware. In the equipment selection process, changes in demand-side technology (such as improvements in the efficiencies of space heating and cooling equipment) and evolving provider needs (such as automation of the utility's distribution system) should be evaluated. In addition, distribution utilities need also to identify the safeguards that will ensure proper equipment use.

Establishing an Appropriate Implementation Program

Some marketing programs are better suited to promote the installation of certain demand-side technology alternatives. In most cases, a mix of implementation techniques will be used. The selected implementation marketing measures should be compatible with any technology alternative that is part of the demand-side program.

Identifying Quality Assurance Considerations

Because of the possible large number of dispersed devices, service providers can improve customer and utility system performance by considering quality assurance in the implementation program. In the example of a direct load control program, failures may be attributed either to malfunctions of the devices or to the communication links.

Developing an Installation and Maintenance Schedule

Many expenses are involved in the installation, maintenance, and repair of the numerous utility- and third-party-owned and operated devices that are included in a demand-side program. Providers can reduce operating costs by developing prudent scheduling policies for their limited crew resources. Efficient scheduling of equipment ordering and installation is helpful in reducing unnecessary program delays.

THE IMPLEMENTATION PROCESS

Developing, installing, and operating a generating plant—that is, all the steps associated with "implementing" a supply-side program—takes years of planning and scheduling, rigorous analytic modeling, calculations concerning reliability and maintenance, and strict construction scheduling. An equally rigorous approach is needed to implement the demand-side alternatives. There are many utility and non-utility actors involved in the implementation process, and this requires the careful coordination of all parties. Figure 13-3 illustrates the typical process.

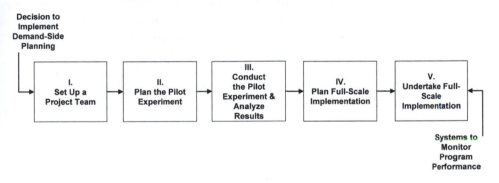

Figure 13-3. Stages in the Implementation Process of Demand-side Programs or Activities

The implementation process takes place in several stages. The stages may include forming an implementation project team, completing a pilot experiment and demonstration, and, finally, expanding to system-side implementation. This "time-phased" process tends to reduce the magnitude of the implementation problem because pilot programs can be used to resolve program problems before system-wide implementation takes place.

Implementing demand-side programs involves almost every functional department within a service provider, and careful coordination is required. As a first step, market participants may want to create a high-level, demand-side planning project team with representation from the various departments, and with the overall control and responsibility for the implementation process. It is important for management to establish clear directives for the project team, including a written scope of responsibility, project team goals and time frame.

When limited information is available on prior demand-side program experiences, a pilot experiment may precede the program. Pilot experiments may be limited either to a sub-region or to a sample of customers. If the pilot experiment proves cost-effective, then initiating the full-scale program may be considered.

After the pilot experiment is completed, additional effort must be given to refining the training, staffing, marketing, and program administration requirements.

How Should Monitoring and Evaluation of the Performance of Demand-side Programs and Activities Be Best Achieved?

Just as there is need to monitor the performance of supply-side alternatives, there is a need to monitor demand-side alternatives.* The ultimate goal of the monitoring program is to identify deviations from expected performance and to improve both existing and planned demand-side programs. Monitoring and evaluation programs can also serve as a primary source of information on customer behavior and system impacts, foster advanced planning and organization within a demand-side program, and provide management with the means of examining demand-side programs as they develop.

In monitoring the performance of demand-side programs, two questions need to be addressed:

- Was the program implemented as planned?
- Did the program achieve its objectives?

The first question may be fairly easy to answer once a routine monitoring system has been adopted. Tracking and review of program costs, customer acceptance, and scheduled milestones can help determine whether the demand-side management program has been implemented as planned.

---

*Portions of the material presented in this section have been adapted from various evaluation studies by Synergic Resources Corporation and by Eric Hirst and colleagues at Oak Ridge National Laboratory.

The second question can be much more difficult to answer. As noted previously, demand-side program objectives can be best characterized in terms of load shape changes. Thus, an assessment of program success must begin with measuring the impact of the program on load shapes. However, this measurement can be difficult because other factors unrelated to the utility's demand-side program can have a significant impact on customer loads.

MONITORING AND EVALUATION APPROACHES

In monitoring and evaluating demand-side programs, two common approaches can be taken:

- **Descriptive**—Basic monitoring that includes documentation of program costs, activities completed, services offered, customer acceptance rates, and characteristics of program participants.

- **Experimental**—Use of comparisons and control groups to determine relative program effects on participants or non-participants, or both.

The two monitoring approaches tend to address different sets of concerns; therefore, it may be useful to incorporate both in programs. With a descriptive approach, management should be aware of basic program performance indicators, in terms of both administrative procedure and target population characteristics. Information such as the cost per unit of service, the frequency of demand-side equipment installation, the type of participants (single family households or other demographic groups), and the number of customer complaints, can be useful in assessing the relative success of a demand-side management program. Recordkeeping and reporting systems can be helpful in completing descriptive evaluations.

The descriptive evaluation, however, is not adequate for systematically assessing the load shape impacts of the demand-side program. To assess load shape impacts requires the careful definition of a reference baseline against which load shapes with a demand-side management alternative can be judged. The reference baseline reflects those load shape changes that are "naturally occurring"—that is, those changes unrelated to the demand-side program itself.

In some cases, the reference baseline might be the existing forecast with appropriate adjustment to reflect the short-term conditions in the service area. In other cases, the reference might be a control group of customers not participating in the program. The energy consumption and hourly demand of program participants could also be measured before they joined the program to provide a "before" and "after" comparison.

If the "before" reference is used, it is often necessary to adjust data for subsequent changes in the customer's appliance or equipment stock and its usage. Adding an extra appliance (such as a window air conditioner) can more than offset any reductions in energy use resulting from a weatherization program. Similarly, the effect of direct load control of a water heater can be altered by changes in a customer's living pattern resulting, for example, from retirement or from the second spouse joining the labor force. The "before" situation must be clearly characterized so that appropriate impact of the program can be measured.

If the reference point is a group of non-participants, that group must have characteristics similar to those of the participants. This, of course, requires a great deal of information on both groups to allow for proper matching, including not only appliance stock data but also such information as family size, work schedules, income, and age of head of household.

ISSUES IN PROGRAM MONITORING AND EVALUATION

Although there are numerous issues facing utility management as it undertakes demand-side program monitoring, most of these issues can be grouped into one of four categories: monitoring program validity, data and information requirements, management concerns, and program organization.

Monitoring Program Validity

Monitoring programs strive to achieve two types of validity: internal and external. Internal validity is the ability to accurately measure the effect of the demand-side management program on the participant group itself. External validity is the ability to generalize experimental results to the entire population. For example, controlled water heating may reduce peak load for a sample of participants, but there is no guarantee that all customers will react in a similar fashion.

Threats to monitoring program validity usually fall into two categories: problems associated with randomization and problems associated with confounding influences. Randomization refers to the degree to which the participating customer sample truly represents the total customer population involved in the demand-side program. It can also refer to the degree of bias involved in assigning customers to the experimental and control groups.

Confounding influences refer to non-program-related changes that may increase or decrease the impact of a demand-side program. In some cases, the effect of these non-program-related changes can be greater than the effect of the demand-side program. The list of potential sources of confounding influences is extensive, including weather, inflation, changes in personal income, and plant openings and closings.

Data and Information Requirements

Data and information requirements involve the entire process of collecting, managing, validating, and analyzing data in the monitoring and evaluation program. The cost of data collection is likely to be the most expensive part of the evaluation study. Data collection costs can be reduced with proper advance planning and by having sufficient recordkeeping and reporting systems. Sources of evaluation data include program records, customer bills, metering, and field surveys. Typically, telephone surveys are used by utilities to complete field surveys. The data must be valid (measure what it is supposed to measure) and reliable (the same results would occur if repeated, within an acceptable margin of error).

The data collection system should be designed before the implementation of the demand-side program itself. There are a number of reasons for this:

- Some information is needed on a "before" and "after" program initiation basis. If the "before" data are inadequate, additional data must be collected before the start of the program.

- Some data collections take considerable time, particularly if metering of customer end uses has to be performed.

- Monitoring the effect of the demand-side alternative throughout the program allows for adjustments and modifications to the program.

In addition, information on participant characteristics, awareness of the program, motivations for participating, and satisfaction are very important in evaluating the overall success of a program.

The information-gathering mechanisms may already be in place at many utilities. Load research programs and customer surveys have long been used to collect data for forecasting and planning. The expertise gained in conducting these activities is helpful in considering the development of a monitoring program.

Management Concerns

Monitoring and evaluation programs require careful management attention. Some of the most important hurdles that must be overcome in the management of a monitoring program include:

- Assuring sufficient advanced planning to develop and implement the monitoring and evaluation program in conjunction with the demand-side activity.

- Recognizing that monitoring and evaluation programs can be data intensive and time consuming; therefore, evaluation program costs must be kept in balance with benefits.

- Establishing clear lines of responsibility and accountability for program formulation and direction.

- Organizing and reporting the results of the evaluation program to provide management with a clear understanding of these programs.

- Developing a strong organizational commitment to adequately plan, coordinate and fund monitoring programs.

MONITORING AND EVALUATION PROGRAMS

Monitoring and evaluation programs can be organized in four stages: pre-evaluation planning, evaluation design, evaluation design implementation, and program feedback.

Pre-Evaluation Planning consists of working out beforehand the conceptual and logistical questions that will be encountered during the full-scale program. The "pre-program planning checklist" presented below will help guide this phase. Once the questions on the checklist have been answered, an overall assessment can be made of the needs of a monitoring and evaluation program. Refer to Table 13-5.

Evaluation Design focuses on organizing the overall evaluation effort and developing of specific program evaluation designs. A decision whether to use descriptive or experimental or some combination of the two must be made. The major elements of an evaluation design are:

- Evaluation Objectives
- Evaluation Approach
- Data Requirements and Collection Strategy
- Data Analysis Procedures
- Final Report Format
- Program Cost

Table 13-5. Pre-program Planning Checklist

Monitoring & Evaluation Program Goals
- **What are the objectives of the demand-side activity, and are they measurable?**
- **Are there secondary objectives to consider?**
- **Are there clear and measurable goals associated with implementing the demand-side activity?**
- **Are the cost implications of demand-side program or activity sufficient to warrant evaluation cost?**
- Can results from the monitoring and evaluation program be used for decision making?
- Are there clear evaluation and monitoring program goals?
- Can an evaluation be completed?

Methods
- What program results are expected?
- Can the results be generalized to a larger population?
- Can a control group or comparative analysis be used?
- What are the alternative evaluation designs?
- What statistical techniques could be used?
- What are the potential sources of bias from confounding factors?
- Will the methods require a significant adaptation to a local conditions?
- Will all survey instruments be pre-tested?

Data Requirements & Processing
- **Are there sufficient program records and other secondary data to support an evaluation?**
- **What are the requirements for new data collection?**
- **How can the data collection and analysis activity best be carried out within the existing data processing system?**

Management & Other Issues
- Who should conduct the evaluations (consultants, internal staff)?
- Are there political and social issues that should be addressed?
- What is the cost of the monitoring and evaluation program?
- What degree of precision is required?
- What is the time frame of analysis?
- Is there sufficient advance time to undertake a suitable monitoring and evaluation program?
- How will responsibility and accountability for the monitoring and evaluation program be assigned?

• Program Management Plan

Design Implementation refers to initiating the evaluation framework according to the action plan, including monitoring, analysis, and recordkeeping.

Program Feedback consists of reviewing the evaluation results and determining whether any program changes are warranted.

How Do I Get Started in Addressing Demand-side Planning Issues as They Relate to My Utility?

The previous discussion has focused on seven major issues related to the assessment of demand-side alternatives for an individual energy service. The information requirements to address these issues are, at first glance, extensive and a major hurdle in getting started in demand-side planning. Typically, the information can be obtained from different organizations, in different geographic locations, for different time periods, and using different collection methods. Thus, some adjustments and generalizations are often required. In any case, there is often a sufficient information base to initiate a preliminary analysis of demand-side alternatives.

In-house information that is of particular use includes:

• Customer appliance surveys.

• Consumer behavior surveys.

• Load research data on customer classes or specific end-use load shapes.

• In-house programs or special studies on specific end uses, such as cooling, electric space heating, or water heating.

• Cost comparisons for major end uses, primarily space heating, cooling, and water heating.

• Information available from sources outside the utility includes:

• Energy consumption data for residential appliances.

• Load shapes for selected appliances and adjustments to make them appropriate to selected service areas.

• Customer responses to time-of-use rates.

- Descriptions and operating characteristics of specific demand-side alternatives.

- Customer acceptance and market penetration of selected demand-side alternatives.

- Analysis tools for the evaluation of demand-side alternatives.

- Housing characteristics and space and water heating system by dwelling type.

References
Demand-side Management Concepts and Methods, C.W. Gellings and J.H. Chamberlin, Fairmont Press, 1993

Appendix

Additional Resources

A Survey Of Time-of-Use (TOU) Pricing and Demand-Response (DR) Programs, prepared for the U.S. Environmental Protection Agency, prepared by Energy & Environmental Economics, San Francisco, CA: July 2006.

Annual Energy Outlook 2006, With Projections to 2030, Energy Information Administration, Office of Integrated Analysis and Forecasting, U.S. Department of Energy, Washington, DC: February 2006.

Climate Change 2007: Mitigation, Contribution of Working Group III to the Fourth Assessment Report of the Intergovernmental Panel on Climate Change, B. Metz, O. R. Davidson, P. R. Bosch, R. Dave, L. A. Meyer (eds), Cambridge University Press, Cambridge, United Kingdom and New York, NY, USA: 2007.

Dynamic Energy Management, EPRI, Palo Alto, CA: 2007

Edahiro, J., "An Overview of Efforts in Japan to Boost Energy Efficiency," *JFS Newsletter,* Sep. 2007.

Ellis, M., International Energy Agency, *Standby Power and the IEA,* Presentation, Berlin, May, 2007.

Energy for Sustainable Development: Energy Policy Options for Africa, UN-ENERGY/Africa: 2007.

Energy Use in the New Millennium: Trends in IEA Countries, International Energy Agency, Paris, France: 2007.

Enge, K., S. Holmen, and M. Sandbakk, *Investment Aid and Contract Bound Energy Savings: Experiences from Norway,* 2007 Summer Study on Energy Efficiency in Industry, White Plains, New York: July 24-27, 2007.

Enova's results and activities in 2006, Enova: 2007. (In Norwegian).

Facilitating the Transition to a Smart Electric Grid, Testimony of Audrey Zibelman, PJM Chief Operating Officer and Executive Vice President, before U.S. House of Representatives Committee on Energy and Commerce, May 3, 2007.

Gellings, C.W., and K.E. Parmenter, "Demand-side Management," in

Handbook of Energy Efficiency and Renewable Energy, edited by F. Kreith and D.Y. Goswami, CRC Press, New York, NY: 2007.

Gellings, C.W., Wikler, G., and Ghosh, D., "Assessment of U.S. Electric End-Use Energy Efficiency Potential," *The Electricity Journal*, Vol. 19, Issue 9, November 2006.

Memorandum of Understanding between the Department of Energy of the United States of America and the National Development and Reform Commission of the People's Republic of China Concerning Industrial Energy Efficiency Cooperation, signed in San Francisco, U.S. Department of Energy, September 12, 2007.

"Merkel confronts German energy industry with radical policy overhaul," Herald Tribune, July 4, 2007.

National Energy Efficiency Best Practices Study, Vol. R1—Residential Lighting Best Practices Report, Quantum Consulting Inc., Berkeley, CA: Dec. 2004.

Price, L., and W. Xuejun, "Constraining Energy Consumption of China's Largest Industrial Enterprises Through the Top-1000 Energy-Consuming Enterprise Program," *2007 Summer Study on Energy Efficiency in Industry*, White Plains, New York, July 24-27, 2007.

Realizing the Potential of Energy Efficiency, Targets, Policies, and Measures for G8 Countries, United Nations Foundation, Washington, DC: 2007.

Standby Power Use and the IEA "1-Watt Plan," Fact Sheet, International Energy Agency, Paris, France: April 2007.

Statistisk Centralbyra, *Statistics Norway, Total supply and use of energy, 1997-2006.*

Stern, N.H., *The Economics of Climate Change: The Stern Review*, Cambridge University Press, Cambridge, UK: 2007.

Taking Action Against Global Warming: An Overview of German Climate Policy, Federal Ministry for the Environment, Nature Conservation and Nuclear Safety, September 2007.

Specker, S., Gellings, C, and Mansoor, A., *The ElectriNet^{SM}: An Electric Power System for a Carbon-Constrained Future, Draft*, Electric Power Research Institute, September 2008.

The Green Grid: How the Smart Grid will Save Energy and Reduce Carbon Emissions, EPRI, Palo Alto, CA and Global Energy Partners, LLC, Lafayette, CA: pending publication.

The Keystone Dialogue on Global Climate Change, Final Report, The Keystone Center, CO: May 2003.

The Power to Reduce CO_2 Emissions: The Full Portfolio, EPRI, Palo Alto,

CA: August 2007.

U.S. and China Sign Agreement to Increase Industrial Energy Efficiency, DOE to Conduct Energy Efficiency Audits on up to 12 Facilities, Press Release, U.S. Department of Energy, September 14, 2007.

"Wal-Mart Continues to Change the Retail World—One CFL at a Time," *Power Tools,* Winter 2006-2007, Vol. 4, No. 4, Global Energy Partners, LLC, Lafayette, CA: 2007.

World Energy Outlook 2006, International Energy Agency, Paris, France: 2006.

World Energy Outlook 2005: Middle East and North Africa Insights, International Energy Agency, Paris, France: 2005.

"Retrofitting Utility Power Plant Motors for Adjustable Speed: Field Test Program," M.J. Samotyj, Program Manager, EPRI Report CU-6914, December 1990.

"Distribution Efficiency Initiative, Market Progress Evaluation Report, No. 1," Global Energy Partners, LLC, report number E05-139, May 18, 2005.

Fetters, John L. "Transformer Efficiency." Electric Power Research Institute. April 23, 2002.

Energy Conservation Program for Commercial Equipment: Distribution Transformers Energy Conservation Standards; Final Rule. 10 CFR Part 431. Federal Register, Vol. 72, No. 197, 58189. October 12, 2007.

Testimony of C.W. Gellings," State of New Jersey—Board of Public Utilities—Appendix II, Group II Load Management Studies, January 1981.

"Demand-side Planning," C.W. Gellings, Edison Electric Institute Executive Symposium for Customer Service and Marketing Personnel, November 1982.

Gellings, Clark W., *Demand-side Management: Volumes 1-5,* EPRI, Palo Alto, CA, 1984-1988.

Assessment of Demand Response and Advanced Metering—Staff Report, FERC Docket AD06-2-000, August 2006.

Dimensions of Demand Response: Capturing Customer Based Resources in New England's Power Systems and Markets—Report and Recommendations of the New England Demand Response Initiative, July 23, 2003.

Arens, E., et al. 2006. "Demand Response Enabling Technology Development, Phase I Report: June 2003-Novembr 2005, Overview and

Reports from the Four Project Groups," Report to CEC Public Interest Energy Research (PIER) Program, Center for the Built Environment, University of California, Berkeley, April 4.

Transmission Control Protocol/Internet Protocol (TCP/IP) as developed by the Advance Research Projects Agency (ARPA) and may be used over Ethernet networks and the Internet. Use of this communications industry standard allows DDC network configurations consisting of off-the-shelf communication devices such as bridges, routers and hubs. Various DDC system manufacturers have incorporated access via the Internet through an IP address specific to the DDC system.

Decision Focus, Inc., Demand-side Planning Cost/Benefit Analysis, November 1983. Published by the Electric Power Research Institute, Report No. EPRI RDS 94 (RP 1613).

A. Faruqui, P.C. Gupta, and J. Wharton, "Ten Propositions in Modeling Industrial Electricity Demand," in Adela Bolet (ed.) Forecasting U.S. Electricity Demand (Boulder, CO, Westview Press), 1985.

Alliance to Save Energy, *Utility Promotion of Investment in Energy Efficiency: Engineering, Legal, and Economic Analyses*, August 1983.

Boston Pacific Company, *Office Productivity Tools for the Information Economy: Possible Effects on Electricity Consumption*, prepared for Electric Power Research Institute, September 1986.

Cambridge Systematics, Inc., Residential End-Use Energy Planning System (REEPS), July 1982. Published by the Electric Power Research Institute, Report No. EA-2512 (RP 1211-2).

"Cooling Commercial Buildings with Off-Peak Power," EPRI Journal, Volume 8, Number 8, October 1983.

Commend building types, the COMMEND Planning System: National and Regional Data and Analysis, EPRI EM-4486.

Customer's Attitudes and Customers' Response to Load Management, Electric Utility Rate Design Study, December 1983. Published by the Electric Power Research Institute, Report No. EPRI SIA82-419-6.

C.W. Gellings and D.R. Limaye, "Market Planning for Electric Utilities," Paper Presented at Energy Technology Conference, Washington, D.C., March 1984.

David C. Hopkins, *The Marketing Plan* (New York: The Conference Board, 1981). Pp. 27-28 and 38-39.

Decision Focus, Inc., Cost/Benefits Analysis of Demand-side Planning Alternatives. Published by Electric Power Research Institute, October 1983, EPRI EURDS 94 (RP 1613).

Decision Focus, Inc. Demand-side Planning Cost/Benefit Analysis, November 1983. Published by the Electric Power Research Institute, Report No. EPRI RDS 94 (RP 1613).

Decision Focus, Inc., Integrated Analysis of Load Shapes and Energy Storage, March 1979. Published by the Electric Power Research Institute. Report No. EA-970 (RP 1108).

Decision Focus, Inc., Load Management Strategy Testing Model. Published by the Electric Power Research Institute, May 1982, EPRI EA 2396 (RP 1485).

Demand-side Management Vol. 1: Overview of Key Issues, EPRI EA/EM-3597, Vol. 1 Project 2381-4 Final Report.

Demand-side Management Vol. 2: Evaluation of Alternatives, EPRI EA/EM-3597, Vol. 2 Project 2381-4 Final Report.

Demand-side Management Vol. 3: Technology Alternatives and Market Implementation Methods, EPRI EA/EM-3597, Vol. 3 Project 2381-4 Final Report.

Demand-side Management Vol. 4: Commercial Markets and Programs, EPRI EA/EM-3597, Vol. 4 Project 2381-4 Final Report.

EBASCO Services, Inc., Survey of Innovative Rate Structures, forthcoming. Electric Power Research Institute, EPRI RP2381-5.

Eco-Energy Associates, Opportunities in Thermal Storage R&D, July 1983. Published by the Electric Power Research Institute, Report No. EM-3159-SR.

Energy Management Associates, Inc., Issues in Implementing a Load Management Program for Direct Load Control, March 1983. Published by the Electric Power Research Institute, Report No. EPRI EA-2904 (RP 2050-8).

Energy Utilization Systems, Inc., 1981 Survey of Utility Load Management Conservation and Solar End-Use Projects, November 1982. Published by Electric Power Research Institute, Report No. EM-2649 (RP 1940-1).

EPRI Project, RP 2547, Consumer Selection of End-Use Devices and Systems.

EPRI Reports prepared by Synergic Resources Corporation. Electric Utility Conservation Programs: Assessment of Implementation Experience (RP 2050-11) and 1983 Survey of Utility End-Use Projects (EPRI Report No. EM 3529).

Gellings, Clark W., Pradeep C. Gupta, and Ahmad Faruqui. Strategic

Implications of Demand-side Planning," in James L. Plummer (ed.) Strategic Planning and Management for Electric Utilities, Prentice-Hall, New Jersey, forthcoming 1984.

Identifying Commercial Industrial Market Segments for Utility Demand-side Programs, Gayle Lloyd, Jersey Central Power and Light Company and Todd Davis. Synergic Resources Corporation, The PG and E Energy Expo, April 1966.

J.S. McMenamin and I. Rohmund, *Electricity Use in the Commercial Sector*: Insights from EPRI Research, Electric Power Research Institute Working Paper, March 1986.

Lauritis R. Chirtensen Associates, Inc., Residential Response to Time-of-Use Rates, EPRI Project RP 1956.

Linda Finley, "Load Management Implementation Issues," December, 1982, presentation made at the EPRI Seminar on "Planning and Assessment of Load Management."

M.A. Kuliasha, Utility Controlled Customer Side Thermal Energy Storage Tests: Cool Storage, February 1983. Published by Oak Ridge National Laboratory, Report No. ORNL-5795.

Marketing Demand-side Programs to Improve Load Factor, Electric Power Research Institute, EA-4267, October 1985, p. 5-4.

Mathematical Sciences Northwest, Inc., Reference Manual of Data Sources for Load Forecasting, September 1981. Published by Electric Power Research Institute, EPRI EA-2008 (RP 1478-1).

Methods for Analyzing the Market Penetration of End-Use Technologies: A Guide for Utility Planners, Published by the Electric Power Research Institute, October 1982, EPRI EA-2702 (RP 2045-2).

Michael Porter, *Competitive Strategy* (New York Free Press, 1980).

Non-residential Buildings Energy Consumption Survey: *Characteristics of Commercial Buildings*, 1983. U.S. Department of Energy, Energy Information Administration, p. 57.

Pradeep Gupta. "Load Forecasting," from the Utility Resource Planning Conference sponsored by the University of California-Berkeley, College of Engineering, February 28, 1984. Berkeley, California.

Resource Planning Associates, Inc. Methods for Analyzing the Market Penetration of End-Use Technologies; A Guide for Utility Planners, published by the Electric Power Research Institute, October 1982, EPRI EA-2702 (RP 2045-2). EPRI has also recently funded a project on "Identifying Consumer Research Techniques for Electric Utilities" (RP 1537).

Robert M. Coughlin, *"Understanding Commercial Fuel and Equipment Choice Decisions." Meeting Energy Challenges: The Great PG&E Energy Expo,* 1985. Conference Proceedings, vol. 2, edited by Craig Smith, Todd Davis and Peter Turnbull (New York Pergamon Press, Inc. 1985), pp. 439-447.

Southern California Edison Company, 1981 Conservation and Load Management: Volume II Measurement (1981 Page 2-VIII-I).

State Energy Date Report, U.S. Department of Energy, April, 1986 Statistical Year Book, Edison Electric Institute, 1985.

Stephen Braithwait, Residential Load Forecasting: Integrating End Use and Econometric Methods. Paper presented at Utility Conservation Programs: Planning, Analysis, and Implementation, New Orleans, September 13, 1983.

Survey of Utility Commercial Sector Activities, EPRI EM-4142, July 1985.

Synergic Resources Corporation, Electric Utility Sponsored Conservation Programs: An Assessment of Implementation Mechanisms (forthcoming), Electric Power Research Institute, RP 2050-11.

Synergic Resources Corporation, 1983 Survey of Utility End-Use Projects, Electric Power Research Institute, Report EM-3529-1984 (RP 1940-8).

U.S. Department of Labor, Bureau of Labor Statistics, US DL 85-478, November 7, 1985.

1979 Nonresidential Buildings, Energy Consumption Survey Data for COMMEND buildings.

1983—1987 Research and Development Program Plan published by the Electric Power Research Institute, January 1983, EPRI P-2799-SR.

Index